LECTURE
LINEAR ALGEBRA

I.M. GEL'FAND

Translated by A. Shenitzer
York University

DOVER PUBLICATIONS, INC.
New York

Published in Canada by General Publishing Company, Ltd., 30 Lesmill Road, Don Mills, Toronto, Ontario.
Published in the United Kingdom by Constable and Company, Ltd., 10 Orange Street, London WC2H 7EG.

This Dover edition, first published in 1989, is an unabridged and unaltered republication of the work first published by Interscience Publishers, Inc., New York, in 1961. The first Russian edition was published in 1948 and the revised second Russian edition, of which this book is a translation, in 1950.

Manufactured in the United States of America
Dover Publications, Inc., 31 East 2nd Street, Mineola, N.Y. 11501

Library of Congress Cataloging-in-Publication Data

Gel'fand, I. M. (Izrail ' Moiseevich)
[Lektsii po lineinoi algebre. English]
Lectures on linear algebra / I. M. Gel'fand ; translated by A. Shenitzer.
p. cm.
Translation of: Lektsii po lineinoi algebre.
Previously published: New York : Interscience Publishers, 1961.
ISBN 0-486-66082-6
1. Algebras, Linear. I. Title.
QA184.G4413 1989
512'.5—dc20 89-35381
CIP

PREFACE TO THE SECOND EDITION

The second edition differs from the first in two ways. Some of the material was substantially revised and new material was added. The major additions include two appendices at the end of the book dealing with computational methods in linear algebra and the theory of perturbations, a section on extremal properties of eigenvalues, and a section on polynomial matrices (§§ 17 and 21). As for major revisions, the chapter dealing with the Jordan canonical form of a linear transformation was entirely rewritten and Chapter IV was reworked. Minor changes and additions were also made. The new text was written in collaboration with Z. Ia. Shapiro.

I wish to thank A. G. Kurosh for making available his lecture notes on tensor algebra. I am grateful to S. V. Fomin for a number of valuable comments. Finally, my thanks go to M. L. Tzeitlin for assistance in the preparation of the manuscript and for a number of suggestions.

September 1950 I. GEL'FAND

Translator's note: Professor Gel'fand asked that the two appendices be left out of the English translation.

PREFACE TO THE FIRST EDITION

This book is based on a course in linear algebra taught by the author in the department of mechanics and mathematics of the Moscow State University and at the Byelorussian State University.

S. V. Fomin participated to a considerable extent in the writing of this book. Without his help this book could not have been written.

The author wishes to thank Assistant Professor A. E. Turetski of the Byelorussian State University, who made available to him notes of the lectures given by the author in 1945, and to D. A. Raikov, who carefully read the manuscript and made a number of valuable comments.

The material in fine print is not utilized in the main part of the text and may be omitted in a first perfunctory reading.

January 1948 I. GEL'FAND

TABLE OF CONTENTS

Page

I. n-Dimensional Spaces. Linear and Bilinear Forms 1
§ 1. n-Dimensional vector spaces 1
§ 2. Euclidean space 14
§ 3. Orthogonal basis. Isomorphism of Euclidean spaces 21
§ 4. Bilinear and quadratic forms 34
§ 5. Reduction of a quadratic form to a sum of squares 42
§ 6. Reduction of a quadratic form by means of a triangular transformation 46
§ 7. The law of inertia 55
§ 8. Complex n-dimensional space 60

II. Linear Transformations 70
§ 9. Linear transformations. Operations on linear transformations . 70
§ 10. Invariant subspaces. Eigenvalues and eigenvectors of a linear transformation. 81
§ 11. The adjoint of a linear transformation. 90
§ 12. Self-adjoint (Hermitian) transformations. Simultaneous reduction of a pair of quadratic forms to a sum of squares 97
§ 13. Unitary transformations 103
§ 14. Commutative linear transformations. Normal transformations . 107
§ 15. Decomposition of a linear transformation into a product of a unitary and self-adjoint transformation 111
§ 16. Linear transformations on a real Euclidean space 114
§ 17. Extremal properties of eigenvalues 126

III. The Canonical Form of an Arbitrary Linear Transformation . 132
§ 18. The canonical form of a linear transformation 132
§ 19. Reduction to canonical form 137
§ 20. Elementary divisors 142
§ 21. Polynomial matrices 149

IV. Introduction to Tensors 164
§ 22. The dual space 164
§ 23. Tensors 171

CHAPTER I

n-Dimensional Spaces. Linear and Bilinear Forms

§ 1. n-Dimensional vector spaces

1. *Definition of a vector space.* We frequently come across objects which are added and multiplied by numbers. Thus

1. In geometry objects of this nature are vectors in three dimensional space, i.e., directed segments. Two directed segments are said to define the same vector if and only if it is possible to translate one of them into the other. It is therefore convenient to measure off all such directed segments beginning with one common point which we shall call the origin. As is well known the sum of two vectors $\mathbf{x}$ and $\mathbf{y}$ is, by definition, the diagonal of the parallelogram with sides $\mathbf{x}$ and $\mathbf{y}$. The definition of multiplication by (real) numbers is equally well known.

2. In algebra we come across systems of n numbers $\mathbf{x} = (\xi_1, \xi_2, \cdots, \xi_n)$ (e.g., rows of a matrix, the set of coefficients of a linear form, etc.). Addition and multiplication of n-tuples by numbers are usually defined as follows: by the sum of the n-tuples $\mathbf{x} = (\xi_1, \xi_2, \cdots, \xi_n)$ and $\mathbf{y} = (\eta_1, \eta_2, \cdots, \eta_n)$ we mean the n-tuple $\mathbf{x} + \mathbf{y} = (\xi_1 + \eta_1, \xi_2 + \eta_2, \cdots, \xi_n + \eta_n)$. By the product of the number λ and the n-tuple $\mathbf{x} = (\xi_1, \xi_2, \cdots, \xi_n)$ we mean the n-tuple $\lambda\mathbf{x} = (\lambda\xi_1, \lambda\xi_2, \cdots, \lambda\xi_n)$.

3. In analysis we define the operations of addition of functions and multiplication of functions by numbers. In the sequel we shall consider all continuous functions defined on some interval $[a, b]$.

In the examples just given the operations of addition and multiplication by numbers are applied to entirely dissimilar objects. To investigate all examples of this nature from a unified point of view we introduce the concept of a vector space.

DEFINITION 1. *A set* $\mathbf{R}$ *of elements* $\mathbf{x}, \mathbf{y}, \mathbf{z}, \cdots$ *is said to be a vector space over a field* F *if:*

(a) *With every two elements* $\mathbf{x}$ *and* $\mathbf{y}$ *in* $\mathbf{R}$ *there is associated an element* $\mathbf{z}$ *in* $\mathbf{R}$ *which is called the sum of the elements* $\mathbf{x}$ *and* $\mathbf{y}$. *The sum of the elements* $\mathbf{x}$ *and* $\mathbf{y}$ *is denoted by* $\mathbf{x} + \mathbf{y}$.

(b) *With every element* $\mathbf{x}$ *in* $\mathbf{R}$ *and every number* λ *belonging to the field F there is associated an element* $\lambda\mathbf{x}$ *in* $\mathbf{R}$. $\lambda\mathbf{x}$ *is referred to as the product of* $\mathbf{x}$ *by* λ.

The above operations must satisfy the following requirements (axioms):

I. 1. $\mathbf{x} + \mathbf{y} = \mathbf{y} + \mathbf{x}$ (commutativity)

2. $(\mathbf{x} + \mathbf{y}) + \mathbf{z} = \mathbf{x} + (\mathbf{y} + \mathbf{z})$ (associativity)

3. $\mathbf{R}$ *contains an element* $\mathbf{0}$ *such that* $\mathbf{x} + \mathbf{0} = \mathbf{x}$ *for all* $\mathbf{x}$ *in* $\mathbf{R}$. $\mathbf{0}$ *is referred to as the zero element.*

4. *For every* $\mathbf{x}$ *in* $\mathbf{R}$ *there exists (in* $\mathbf{R}$*) an element denoted by* $-\mathbf{x}$ *with the property* $\mathbf{x} + (-\mathbf{x}) = \mathbf{0}$.

II. 1. $1 \cdot \mathbf{x} = \mathbf{x}$

2. $\alpha(\beta\mathbf{x}) = \alpha\beta(\mathbf{x})$.

III. 1. $(\alpha + \beta)\mathbf{x} = \alpha\mathbf{x} + \beta\mathbf{x}$

2. $\alpha(\mathbf{x} + \mathbf{y}) = \alpha\mathbf{x} + \alpha\mathbf{y}$.

It is not an oversight on our part that we have not specified how elements of $\mathbf{R}$ are to be added and multiplied by numbers. Any definitions of these operations are acceptable as long as the axioms listed above are satisfied. Whenever this is the case we are dealing with an instance of a vector space.

We leave it to the reader to verify that the examples *1*, *2*, *3* above are indeed examples of vector spaces.

Let us give a few more examples of vector spaces.

4. The set of all polynomials of degree not exceeding some natural number n constitutes a vector space if addition of polynomials and multiplication of polynomials by numbers are defined in the usual manner.

We observe that under the usual operations of addition and multiplication by numbers the set of polynomials of degree n does not form a vector space since the sum of two polynomials of degree n may turn out to be a polynomial of degree smaller than n. Thus

$$(t^n + t) + (-t^n + t) = 2t.$$

5. We take as the elements of $\mathbf{R}$ matrices of order n. As the sum of the matrices $||a_{ik}||$ and $||b_{ik}||$ we take the matrix $||a_{ik} + b_{ik}||$. As the product of the number λ and the matrix $||a_{ik}||$ we take the matrix $||\lambda a_{ik}||$. It is easy to see that the above set **R** is now a vector space.

It is natural to call the elements of a vector space *vectors*. The fact that this term was used in Example *1* should not confuse the reader. The geometric considerations associated with this word will help us clarify and even predict a number of results.

If the numbers $\lambda, \mu, \cdots$ involved in the definition of a vector space are real, then the space is referred to as a *real vector space*. If the numbers $\lambda, \mu, \cdots$ are taken from the field of complex numbers, then the space is referred to as a *complex vector space*.

More generally it may be assumed that $\lambda, \mu, \cdots$, are elements of an arbitrary field K. Then **R** is called a vector space over the field K. Many concepts and theorems dealt with in the sequel and, in particular, the contents of this section apply to vector spaces over arbitrary fields. However, in chapter I we shall ordinarily assume that **R** is a real vector space.

2. *The dimensionality of a vector space.* We now define the notions of linear dependence and independence of vectors which are of fundamental importance in all that follows.

DEFINITION 2. *Let* **R** *be a vector space. We shall say that the vectors* $\mathbf{x}, \mathbf{y}, \mathbf{z}, \cdots, \mathbf{v}$ *are linearly dependent if there exist numbers* $\alpha, \beta, \gamma, \cdots \theta$, *not all equal to zero such that*

$$\alpha\mathbf{x} + \beta\mathbf{y} + \gamma\mathbf{z} + \cdots + \theta\mathbf{v} = \mathbf{0}. \tag{1}$$

Vectors which are not linearly dependent are said to be linearly independent. In other words,

a set of vectors $\mathbf{x}, \mathbf{y}, \mathbf{z}, \cdots, \mathbf{v}$ *is said to be linearly independent if the equality*

$$\alpha\mathbf{x} + \beta\mathbf{y} + \gamma\mathbf{z} + \cdots + \theta\mathbf{v} = \mathbf{0}$$

implies that $\alpha = \beta = \gamma = \cdots = \theta = 0$.

Let the vectors $\mathbf{x}, \mathbf{y}, \mathbf{z}, \cdots, \mathbf{v}$ be linearly dependent, i.e., let $\mathbf{x}, \mathbf{y}, \mathbf{z}, \cdots, \mathbf{v}$ be connected by a relation of the form (1) with at least one of the coefficients, α, say, unequal to zero. Then

$$\alpha\mathbf{x} = -\beta\mathbf{y} - \gamma\mathbf{z} - \cdots - \theta\mathbf{v}.$$

Dividing by α and putting

$$-(\beta/\alpha) = \lambda, \ -(\gamma/\alpha) = \mu, \cdots, -(\theta/\alpha) = \zeta,$$

we have

$$\mathbf{x} = \lambda \mathbf{y} + \mu \mathbf{z} + \cdots + \zeta \mathbf{v}. \tag{2}$$

Whenever a vector $\mathbf{x}$ is expressible through vectors $\mathbf{y}, \mathbf{z}, \cdots, \mathbf{v}$ in the form (2) we say that $\mathbf{x}$ is a *linear combination* of the vectors $\mathbf{y}, \mathbf{z}, \cdots, \mathbf{v}$.

Thus, *if the vectors* $\mathbf{x}, \mathbf{y}, \mathbf{z}, \cdots, \mathbf{v}$ *are linearly dependent then at least one of them is a linear combination of the others.* We leave it to the reader to prove that the converse is also true, i.e., that *if one of a set of vectors is a linear combination of the remaining vectors then the vectors of the set are linearly dependent.*

EXERCISES. *1.* Show that if one of the vectors $\mathbf{x}, \mathbf{y}, \mathbf{z}, \cdots, \mathbf{v}$ is the zero vector then these vectors are linearly dependent.

2. Show that if the vectors $\mathbf{x}, \mathbf{y}, \mathbf{z}, \cdots$ are linearly dependent and $\mathbf{u}, \mathbf{v}, \cdots$ are arbitrary vectors then the vectors $\mathbf{x}, \mathbf{y}, \mathbf{z}, \cdots, \mathbf{u}, \mathbf{v}, \cdots$ are linearly dependent.

We now introduce the concept of *dimension* of a vector space.

Any two vectors on a line are proportional, i.e., linearly dependent. In the plane we can find two linearly independent vectors but any three vectors are linearly dependent. If $\mathbf{R}$ is the set of vectors in three-dimensional space, then it is possible to find three linearly independent vectors but any four vectors are linearly dependent.

As we see the maximal number of linearly independent vectors on a straight line, in the plane, and in three-dimensional space coincides with what is called in geometry the dimensionality of the line, plane, and space, respectively. It is therefore natural to make the following general definition.

DEFINITION 3. *A vector space* $\mathbf{R}$ *is said to be n-dimensional if it contains* n *linearly independent vectors and if any* $n + 1$ *vectors in* $\mathbf{R}$ *are linearly dependent.*

If $\mathbf{R}$ is a vector space which contains an arbitrarily large number of linearly independent vectors, then $\mathbf{R}$ is said to be *infinite-dimensional.*

Infinite-dimensional spaces will not be studied in this book.

We shall now compute the dimensionality of each of the vector spaces considered in the Examples *1, 2, 3, 4, 5.*

1. As we have already indicated, the space **R** of Example *1* contains three linearly independent vectors and any four vectors in it are linearly dependent. Consequently **R** is three-dimensional.

2. Let **R** denote the space whose elements are n-tuples of real numbers.

This space contains n linearly independent vectors. For instance, the vectors

$$\begin{aligned} \mathbf{x}_1 &= (1, 0, \cdots, 0), \\ \mathbf{x}_2 &= (0, 1, \cdots, 0), \\ &\cdots\cdots\cdots\cdots \\ \mathbf{x}_n &= (0, 0, \cdots, 1) \end{aligned}$$

are easily seen to be linearly independent. On the other hand, any m vectors in **R**, $m > n$, are linearly dependent. Indeed, let

$$\begin{aligned} \mathbf{y}_1 &= (\eta_{11}, \eta_{12}, \cdots, \eta_{1n}), \\ \mathbf{y}_2 &= (\eta_{21}, \eta_{22}, \cdots, \eta_{2n}), \\ &\cdots\cdots\cdots\cdots \\ \mathbf{y}_m &= (\eta_{m1}, \eta_{m2}, \cdots, \eta_{mn}) \end{aligned}$$

be m vectors and let $m > n$. The number of linearly independent rows in the matrix

$$\begin{bmatrix} \eta_{11}, & \eta_{12}, & \cdots & \eta_{1n} \\ \eta_{21}, & \eta_{22}, & \cdots & \eta_{2n} \\ \cdots & \cdots & \cdots & \cdots \\ \eta_{m1}, & \eta_{m2}, & \cdots & \eta_{mn} \end{bmatrix}$$

cannot exceed n (the number of columns). Since $m > n$, our m rows are linearly dependent. But this implies the linear dependence of the vectors $\mathbf{y}_1, \mathbf{y}_2, \cdots, \mathbf{y}_m$.

Thus the dimension of **R** is n.

3. Let **R** be the space of continuous functions. Let N be any natural number. Then the functions: $f_1(t) \equiv 1$, $f_2(t) = t, \cdots$, $f_N(t) = t^{N-1}$ form a set of linearly independent vectors (the proof of this statement is left to the reader). It follows that our space contains an arbitrarily large number of linearly independent functions or, briefly, **R** is infinite-dimensional.

4. Let **R** be the space of polynomials of degree $\leqq n - 1$. In this space the n polynomials $1, t, \cdots, t^{n-1}$ are linearly independent. It can be shown that any m elements of **R**, $m > n$, are linearly dependent. Hence **R** is n-dimensional.

5. We leave it to the reader to prove that the space of $n \times n$ matrices $||a_{ik}||$ is n^2-dimensional.

3. *Basis and coordinates in n-dimensional space*

DEFINITION 4. *Any set of n linearly independent vectors $\mathbf{e}_1, \mathbf{e}_2, \cdots, \mathbf{e}_n$ of an n-dimensional vector space* **R** *is called a basis of* **R**.

Thus, for instance, in the case of the space considered in Example *1* any three vectors which are not coplanar form a basis.

By definition of the term "n-dimensional vector space" such a space contains n linearly independent vectors, i.e., it contains a basis.

THEOREM 1. *Every vector* **x** *belonging to an n-dimensional vector space* **R** *can be uniquely represented as a linear combination of basis vectors.*

Proof: Let $\mathbf{e}_1, \mathbf{e}_2, \cdots, \mathbf{e}_n$ be a basis in **R**. Let **x** be an arbitrary vector in **R**. The set $\mathbf{x}, \mathbf{e}_1, \mathbf{e}_2, \cdots, \mathbf{e}_n$ contains $n + 1$ vectors. It follows from the definition of an n-dimensional vector space that these vectors are linearly dependent, i.e., that there exist $n + 1$ numbers $\alpha_0, \alpha_1, \cdots, \alpha_n$ not all zero such that

$$\alpha_0 \mathbf{x} + \alpha_1 \mathbf{e}_1 + \cdots + \alpha_n \mathbf{e}_n = \mathbf{0}. \tag{3}$$

Obviously $\alpha_0 \neq 0$. Otherwise (3) would imply the linear dependence of the vectors $\mathbf{e}_1, \mathbf{e}_2, \cdots, \mathbf{e}_n$. Using (3) we have

$$\mathbf{x} = -\frac{\alpha_1}{\alpha_0}\mathbf{e}_1 - \frac{\alpha_2}{\alpha_0}\mathbf{e}_2 - \cdots - \frac{\alpha_n}{\alpha_0}\mathbf{e}_n.$$

This proves that every $\mathbf{x} \in \mathbf{R}$ is indeed a linear combination of the vectors $\mathbf{e}_1, \mathbf{e}_2, \cdots, \mathbf{e}_n$.

To prove uniqueness of the representation of **x** in terms of the basis vectors we assume that

$$\mathbf{x} = \xi_1 \mathbf{e}_1 + \xi_2 \mathbf{e}_2 + \cdots + \xi_n \mathbf{e}_n$$

and

$$\mathbf{x} = \xi'_1 \mathbf{e}_1 + \xi'_2 \mathbf{e}_2 + \cdots + \xi'_n \mathbf{e}_n.$$

Subtracting one equation from the other we obtain

$$\mathbf{0} = (\xi_1 - \xi'_1)\mathbf{e}_1 + (\xi_2 - \xi'_2)\mathbf{e}_2 + \cdots + (\xi_n - \xi'_n)\mathbf{e}_n.$$

Since $\mathbf{e}_1, \mathbf{e}_2, \cdots, \mathbf{e}_n$ are linearly independent, it follows that

$$\xi_1 - \xi'_1 = \xi_2 - \xi'_2 = \cdots = \xi_n - \xi'_n = 0,$$

i.e.,

$$\xi_1 = \xi'_1, \qquad \xi_2 = \xi'_2, \qquad \cdots, \qquad \xi_n = \xi'_n.$$

This proves uniqueness of the representation.

DEFINITION 5. *If* $\mathbf{e}_1, \mathbf{e}_2, \cdots, \mathbf{e}_n$ *form a basis in an n-dimensional space and*

$$\mathbf{x} = \xi_1\mathbf{e}_1 + \xi_2\mathbf{e}_2 + \cdots + \xi_n\mathbf{e}_n, \tag{4}$$

then the numbers $\xi_1, \xi_2, \cdots, \xi_n$ *are called the coordinates of the vector* $\mathbf{x}$ *relative to the basis* $\mathbf{e}_1, \mathbf{e}_2, \cdots, \mathbf{e}_n$.

Theorem 1 states that *given a basis* $\mathbf{e}_1, \mathbf{e}_2, \cdots, \mathbf{e}_n$ *of a vector space* $\mathbf{R}$ *every vector* $\mathbf{x} \in \mathbf{R}$ *has a unique set of coordinates.*

If the coordinates of $\mathbf{x}$ relative to the basis $\mathbf{e}_1, \mathbf{e}_2, \cdots, \mathbf{e}_n$ are $\xi_1, \xi_2, \cdots, \xi_n$ and the coordinates of $\mathbf{y}$ relative to the same basis are $\eta_1, \eta_2, \cdots \eta_n$, i.e., if

$$\mathbf{x} = \xi_1\mathbf{e}_1 + \xi_2\mathbf{e}_2 + \cdots + \xi_n\mathbf{e}_n$$
$$\mathbf{y} = \eta_1\mathbf{e}_1 + \eta_2\mathbf{e}_2 + \cdots + \eta_n\mathbf{e}_n,$$

then

$$\mathbf{x} + \mathbf{y} = (\xi_1 + \eta_1)\mathbf{e}_1 + (\xi_2 + \eta_2)\mathbf{e}_2 + \cdots + (\xi_n + \eta_n)\mathbf{e}_n,$$

i.e., the coordinates of $\mathbf{x} + \mathbf{y}$ are $\xi_1 + \eta_1, \xi_2 + \eta_2, \cdots, \xi_n + \eta_n$. Similarly the vector $\lambda\mathbf{x}$ has as coordinates the numbers $\lambda\xi_1, \lambda\xi_2, \cdots, \lambda\xi_n$.

Thus *the coordinates of the sum of two vectors are the sums of the appropriate coordinates of the summands, and the coordinates of the product of a vector by a scalar are the products of the coordinates of that vector by the scalar in question.*

It is clear that the zero vector is the only vector all of whose coordinates are zero.

EXAMPLES. *1.* In the case of three-dimensional space our definition of the coordinates of a vector coincides with the definition of the coordinates of a vector in a (not necessarily Cartesian) coordinate system.

2. Let $\mathbf{R}$ be the space of n-tuples of numbers. Let us choose as basis the vectors

$$\begin{aligned} \mathbf{e}_1 &= (1, 1, 1, \cdots, 1), \\ \mathbf{e}_2 &= (0, 1, 1, \cdots, 1), \\ &\cdots\cdots\cdots\cdots\cdots \\ \mathbf{e}_n &= (0, 0, 0, \cdots, 1), \end{aligned}$$

and then compute the coordinates $\eta_1, \eta_2, \cdots, \eta_n$ of the vector $\mathbf{x} = (\xi_1, \xi_2, \cdots, \xi_n)$ relative to the basis $\mathbf{e}_1, \mathbf{e}_2, \cdots, \mathbf{e}_n$. By definition

$$\mathbf{x} = \eta_1 \mathbf{e}_1 + \eta_2 \mathbf{e}_2 + \cdots + \eta_n \mathbf{e}_n;$$

i.e.,

$$\begin{aligned} (\xi_1, \xi_2, \cdots, \xi_n) &= \eta_1(1, 1, \cdots, 1) \\ &+ \eta_2(0, 1, \cdots, 1) \\ &+ \cdots\cdots\cdots\cdots \\ &+ \eta_n(0, 0, \cdots, 1) \\ &= (\eta_1, \eta_1 + \eta_2, \cdots, \eta_1 + \eta_2 + \cdots + \eta_n). \end{aligned}$$

The numbers $(\eta_1, \eta_2, \cdots, \eta_n)$ must satisfy the relations

$$\begin{aligned} \eta_1 &= \xi_1, \\ \eta_1 + \eta_2 &= \xi_2, \\ &\cdots\cdots\cdots \\ \eta_1 + \eta_2 + \cdots + \eta_n &= \xi_n. \end{aligned}$$

Consequently,

$$\eta_1 = \xi_1, \qquad \eta_2 = \xi_2 - \xi_1, \qquad \cdots, \qquad \eta_n = \xi_n - \xi_{n-1}.$$

Let us now consider a basis for **R** in which the connection between the coordinates of a vector $\mathbf{x} = (\xi_1, \xi_2, \cdots, \xi_n)$ and the numbers $\xi_1, \xi_2, \cdots, \xi_n$ which define the vector is particularly simple. Thus, let

$$\begin{aligned} \mathbf{e}_1 &= (1, 0, \cdots, 0), \\ \mathbf{e}_2 &= (0, 1, \cdots, 0), \\ &\cdots\cdots\cdots\cdots \\ \mathbf{e}_n &= (0, 0, \cdots, 1). \end{aligned}$$

Then

$$\begin{aligned} \mathbf{x} &= (\xi_1, \xi_2, \cdots, \xi_n) \\ &= \xi_1(1, 0, \cdots, 0) + \xi_2(0, 1, \cdots, 0) + \cdots + \xi_n(0, 0, \cdots, 1) \\ &= \xi_1 \mathbf{e}_1 + \xi_2 \mathbf{e}_2 + \cdots + \xi_n \mathbf{e}_n. \end{aligned}$$

It follows that *in the space* **R** *of n-tuples* $(\xi_1, \xi_2, \cdots, \xi_n)$ *the numbers*

$\xi_1, \xi_2, \cdots, \xi_n$ *may be viewed as the coordinates of the vector* $\mathbf{x} = (\xi_1, \xi_2, \cdots, \xi_n)$ *relative to the basis*

$$\mathbf{e}_1 = (1, 0, \cdots, 0), \quad \mathbf{e}_2 = (0, 1, \cdots, 0), \quad \cdots, \quad \mathbf{e}_n = (0, 0, \cdots, 1).$$

EXERCISE. Show that in an arbitrary basis

$$\begin{aligned} \mathbf{e}_1 &= (a_{11}, a_{12}, \cdots, a_{1n}), \\ \mathbf{e}_2 &= (a_{21}, a_{22}, \cdots, a_{2n}), \\ &\cdots\cdots\cdots\cdots\cdots \\ \mathbf{e}_n &= (a_{n1}, a_{n2}, \cdots, a_{nn}) \end{aligned}$$

the coordinates $\eta_1, \eta_2, \cdots, \eta_n$ of a vector $\mathbf{x} = (\xi_1, \xi_2, \cdots, \xi_n)$ are linear combinations of the numbers $\xi_1, \xi_2, \cdots, \xi_n$.

3. Let $\mathbf{R}$ be the vector space of polynomials of degree $\leqq n - 1$. A very simple basis in this space is the basis whose elements are the vectors $\mathbf{e}_1 = 1$, $\mathbf{e}_2 = t, \cdots, \mathbf{e}_n = t^{n-1}$. It is easy to see that the coordinates of the polynomial $P(t) = a_0 t^{n-1} + a_1 t^{n-2} + \cdots + a_{n-1}$ in this basis are the coefficients $a_{n-1}, a_{n-2}, \cdots, a_0$.

Let us now select another basis for $\mathbf{R}$:

$$\mathbf{e}'_1 = 1, \quad \mathbf{e}'_2 = t - a, \quad \mathbf{e}'_3 = (t-a)^2, \quad \cdots, \quad \mathbf{e}'_n = (t-a)^{n-1}.$$

Expanding $P(t)$ in powers of $(t - a)$ we find that

$$P(t) = P(a) + P'(a)(t-a) + \cdots + [P^{(n-1)}(a)/(n-1)!](t-a)^{n-1}.$$

Thus the coordinates of $P(t)$ in this basis are

$$P(a), P'(a), \cdots, [P^{(n-1)}(a)/(n-1)!].$$

4. *Isomorphism of n-dimensional vector spaces.* In the examples considered above some of the spaces are identical with others when it comes to the properties we have investigated so far. One instance of this type is supplied by the ordinary three-dimensional space $\mathbf{R}$ considered in Example *1* and the space $\mathbf{R}'$ whose elements are triples of real numbers. Indeed, once a basis has been selected in $\mathbf{R}$ we can associate with a vector in $\mathbf{R}$ its coordinates relative to that basis; i.e., we can associate with a vector in $\mathbf{R}$ a vector in $\mathbf{R}'$. When vectors are added their coordinates are added. When a vector is multiplied by a scalar all of its coordinates are multiplied by that scalar. This implies a parallelism between the geometric properties of $\mathbf{R}$ and appropriate properties of $\mathbf{R}'$.

We shall now formulate precisely the notion of "sameness" or of "isomorphism" of vector spaces.

DEFINITION 6. *Two vector spaces* $\mathbf{R}$ *and* $\mathbf{R}^1$, *are said to be isomorphic if it is possible to establish a one-to-one correspondence* $\mathbf{x} \leftrightarrow \mathbf{x}'$ *between the elements* $\mathbf{x} \in \mathbf{R}$ *and* $\mathbf{x}' \in \mathbf{R}'$ *such that if* $\mathbf{x} \leftrightarrow \mathbf{x}'$ *and* $\mathbf{y} \leftrightarrow \mathbf{y}'$, *then*

1. *the vector which this correspondence associates with* $\mathbf{x} + \mathbf{y}$ *is* $\mathbf{x}' + \mathbf{y}'$,
2. *the vector which this correspondence associates with* $\lambda\mathbf{x}$ *is* $\lambda\mathbf{x}'$.

There arises the question as to which vector spaces are isomorphic and which are not.

Two vector spaces of different dimensions are certainly not isomorphic.

Indeed, let us assume that $\mathbf{R}$ and $\mathbf{R}'$ are isomorphic. If $\mathbf{x}, \mathbf{y}, \cdots$ are vectors in $\mathbf{R}$ and $\mathbf{x}', \mathbf{y}', \cdots$ are their counterparts in $\mathbf{R}'$ then — in view of conditions 1 and 2 of the definition of isomorphism — the equation $\lambda\mathbf{x} + \mu\mathbf{y} + \cdots = \mathbf{0}$ is equivalent to the equation $\lambda\mathbf{x}' + \mu\mathbf{y}' + \cdots = \mathbf{0}$. Hence the counterparts in $\mathbf{R}'$ of linearly independent vectors in $\mathbf{R}$ are also linearly independent and conversely. Therefore the maximal number of linearly independent vectors in $\mathbf{R}$ is the same as the maximal number of linearly independent vectors in $\mathbf{R}'$. This is the same as saying that the dimensions of $\mathbf{R}$ and $\mathbf{R}'$ are the same. It follows that two spaces of different dimensions cannot be isomorphic.

THEOREM 2. *All vector spaces of dimension n are isomorphic.*

Proof: Let $\mathbf{R}$ and $\mathbf{R}'$ be two n-dimensional vector spaces. Let $\mathbf{e}_1, \mathbf{e}_2, \cdots, \mathbf{e}_n$ be a basis in $\mathbf{R}$ and let $\mathbf{e}'_1, \mathbf{e}'_2, \cdots, \mathbf{e}'_n$ be a basis in $\mathbf{R}'$. We shall associate with the vector

$$\mathbf{x} = \xi_1\mathbf{e}_1 + \xi_2\mathbf{e}_2 + \cdots + \xi_n\mathbf{e}_n \tag{5}$$

the vector

$$\mathbf{x}' = \xi_1\mathbf{e}'_1 + \xi_2\mathbf{e}'_2 + \cdots + \xi_n\mathbf{e}'_n,$$

i.e., a linear combination of the vectors $\mathbf{e}'_i$ with the same coefficients as in (5).

This correspondence is one-to-one. Indeed, every vector $\mathbf{x} \in \mathbf{R}$ has a unique representation of the form (5). This means that the ξ_i are uniquely determined by the vector $\mathbf{x}$. But then $\mathbf{x}'$ is likewise uniquely determined by $\mathbf{x}$. By the same token every $\mathbf{x}' \in \mathbf{R}'$ determines one and only one vector $\mathbf{x} \in \mathbf{R}$.

It should now be obvious that if $\mathbf{x} \leftrightarrow \mathbf{x}'$ and $\mathbf{y} \leftrightarrow \mathbf{y}'$, then $\mathbf{x} + \mathbf{y} \leftrightarrow \mathbf{x}' + \mathbf{y}'$ and $\lambda\mathbf{x} \leftrightarrow \lambda\mathbf{x}'$. This completes the proof of the isomorphism of the spaces $\mathbf{R}$ and $\mathbf{R}'$.

In § 3 we shall have another opportunity to explore the concept of isomorphism.

5. *Subspaces of a vector space*

DEFINITION 7. *A subset* $\mathbf{R}'$, *of a vector space* $\mathbf{R}$ *is called a subspace of* $\mathbf{R}$ *if it forms a vector space under the operations of addition and scalar multiplication introduced in* $\mathbf{R}$.

In other words, a set $\mathbf{R}'$ of vectors $\mathbf{x}, \mathbf{y}, \cdots$ in $\mathbf{R}$ is called a subspace of $\mathbf{R}$ if $\mathbf{x} \in \mathbf{R}'$, $\mathbf{y} \in \mathbf{R}'$ implies $\mathbf{x} + \mathbf{y} \in \mathbf{R}'$, $\lambda\, \mathbf{x} \in \mathbf{R}'$.

EXAMPLES. *1.* The zero or null element of $\mathbf{R}$ forms a subspace of $\mathbf{R}$.

2. The whole space $\mathbf{R}$ forms a subspace of $\mathbf{R}$.

The null space and the whole space are usually referred to as *improper* subspaces. We now give a few examples of non-trivial subspaces.

3. Let $\mathbf{R}$ be the ordinary three-dimensional space. Consider any plane in $\mathbf{R}$ going through the origin. The totality $\mathbf{R}'$ of vectors in that plane form a subspace of $\mathbf{R}$.

4. In the vector space of n-tuples of numbers all vectors $\mathbf{x} = (\xi_1, \xi_2, \cdots, \xi_n)$ for which $\xi_1 = 0$ form a subspace. More generally, all vectors $\mathbf{x} = (\xi_1, \xi_2, \cdots, \xi_n)$ such that

$$a_1\xi_1 + a_2\xi_2 + \cdots + a_n\xi_n = 0,$$

where $a_1, a_2, \cdots, a_n$ are arbitrary but fixed numbers, form a subspace.

5. The totality of polynomials of degree $\leqq n$ form a subspace of the vector space of all continuous functions.

It is clear that every subspace $\mathbf{R}'$ of a vector space $\mathbf{R}$ must contain the zero element of $\mathbf{R}$.

Since a subspace of a vector space is a vector space in its own right we can speak of a basis of a subspace as well as of its dimensionality. It is clear that *the dimension of an arbitrary subspace of a vector space does not exceed the dimension of that vector space.*

EXERCISE. Show that if the dimension of a subspace $\mathbf{R}'$ of a vector space $\mathbf{R}$ is the same as the dimension of $\mathbf{R}$, then $\mathbf{R}'$ coincides with $\mathbf{R}$.

A general method for constructing subspaces of a vector space $\mathbf{R}$ is implied by the observation that *if* $\mathbf{e}$, $\mathbf{f}$, $\mathbf{g}$, $\cdots$ *are a (finite or infinite) set of vectors belonging to* $\mathbf{R}$, *then the set* $\mathbf{R}'$ *of all (finite) linear combinations of the vectors* $\mathbf{e}$, $\mathbf{f}$, $\mathbf{g}$, $\cdots$ *forms a subspace* $\mathbf{R}'$ *of* $\mathbf{R}$. The subspace $\mathbf{R}'$ is referred to as *the subspace generated by the vectors* $\mathbf{e}$, $\mathbf{f}$, $\mathbf{g}$, $\cdots$. This subspace is the smallest subspace of $\mathbf{R}$ containing the vectors $\mathbf{e}$, $\mathbf{f}$, $\mathbf{g}$, $\cdots$.

The subspace $\mathbf{R}'$ *generated by the linearly independent vectors* $\mathbf{e}_1, \mathbf{e}_2, \cdots, \mathbf{e}_k$ *is k-dimensional and the vectors* $\mathbf{e}_1, \mathbf{e}_2, \cdots, \mathbf{e}_k$ *form a basis of* $\mathbf{R}'$. Indeed, $\mathbf{R}'$ contains k linearly independent vectors (i.e., the vectors $\mathbf{e}_1, \mathbf{e}_2, \cdots \mathbf{e}_k$). On the other hand, let $\mathbf{x}_1$, $\mathbf{x}_2, \cdots, \mathbf{x}_l$ be l vectors in $\mathbf{R}'$ and let $l > k$. If

$$\begin{aligned}
\mathbf{x}_1 &= \xi_{11}\mathbf{e}_1 + \xi_{12}\mathbf{e}_2 + \cdots + \xi_{1k}\mathbf{e}_k, \\
\mathbf{x}_2 &= \xi_{21}\mathbf{e}_1 + \xi_{22}\mathbf{e}_2 + \cdots + \xi_{2k}\mathbf{e}_k, \\
&\cdots\cdots\cdots\cdots\cdots\cdots\cdots\cdots \\
\mathbf{x}_l &= \xi_{l1}\mathbf{e}_1 + \xi_{l2}\mathbf{e}_2 + \cdots + \xi_{lk}\mathbf{e}_k,
\end{aligned}$$

then the l rows in the matrix

$$\begin{bmatrix}
\xi_{11}, \xi_{12}, \cdots, \xi_{1k} \\
\xi_{21}, \xi_{22}, \cdots, \xi_{2k} \\
\cdots\cdots\cdots\cdots \\
\xi_{l1}, \xi_{l2}, \cdots, \xi_{lk}
\end{bmatrix}$$

must be linearly dependent. But this implies (cf. Example *2*, page 5) the linear dependence of the vectors $\mathbf{x}_1, \mathbf{x}_2, \cdots, \mathbf{x}_l$. Thus the maximal number of linearly independent vectors in $\mathbf{R}'$, i.e., the dimension of $\mathbf{R}'$, is k and the vectors $\mathbf{e}_1, \mathbf{e}_2, \cdots, \mathbf{e}_k$ form a basis in $\mathbf{R}'$.

EXERCISE. Show that every n-dimensional vector space contains subspaces of dimension l, $l = 1, 2, \cdots, n$.

If we ignore null spaces, then the simplest vector spaces are one-dimensional vector spaces. A basis of such a space is a single vector $\mathbf{e}_1 \neq \mathbf{0}$. Thus a one-dimensional vector space consists of all vectors $\alpha\mathbf{e}_1$, where α is an arbitrary scalar.

Consider the set of vectors of the form $\mathbf{x} = \mathbf{x}_0 + \alpha\mathbf{e}_1$, where $\mathbf{x}_0$ and $\mathbf{e}_1 \neq \mathbf{0}$ are fixed vectors and α ranges over all scalars. It is natural to call this set of vectors — by analogy with three-dimensional space — a *line* in the vector space $\mathbf{R}$.

Similarly, all vectors of the form $\alpha\mathbf{e}_1 + \beta\mathbf{e}_2$, where $\mathbf{e}_1$ and $\mathbf{e}_2$ are fixed linearly independent vectors and α and β are arbitrary numbers form a two-dimensional vector space. The set of vectors of the form

$$\mathbf{x} = \mathbf{x}_0 + \alpha\mathbf{e}_1 + \beta\mathbf{e}_2,$$

where $\mathbf{x}_0$ is a fixed vector, is called a (two-dimensional) *plane*.

EXERCISES. *1.* Show that in the vector space of n-tuples $(\xi_1, \xi_2, \cdots, \xi_n)$ of real numbers the set of vectors satisfying the relation

$$a_1\xi_1 + a_2\xi_2 + \cdots + a_n\xi_n = 0$$

$(a_1, a_2, \cdots, a_n$ are fixed numbers not all of which are zero) form a subspace of dimension $n - 1$.

2. Show that if two subspaces $\mathbf{R}_1$ and $\mathbf{R}_2$ of a vector space $\mathbf{R}$ have only the null vector in common then the sum of their dimensions does not exceed the dimension of $\mathbf{R}$.

3. Show that the dimension of the subspace generated by the vectors $\mathbf{e}, \mathbf{f}, \mathbf{g}, \cdots$ is equal to the maximal number of linearly independent vectors among the vectors $\mathbf{e}, \mathbf{f}, \mathbf{g}, \cdots$.

6. *Transformation of coordinates under change of basis.* Let $\mathbf{e}_1, \mathbf{e}_2, \cdots, \mathbf{e}_n$ and $\mathbf{e}'_1, \mathbf{e}'_2, \cdots, \mathbf{e}'_n$ be two bases of an n-dimensional vector space. Further, let the connection between them be given by the equations

$$\begin{aligned}
\mathbf{e}'_1 &= a_{11}\mathbf{e}_1 + a_{21}\mathbf{e}_2 + \cdots + a_{n1}\mathbf{e}_n,\\
\mathbf{e}'_2 &= a_{12}\mathbf{e}_1 + a_{22}\mathbf{e}_2 + \cdots + a_{n2}\mathbf{e}_n,\\
&\cdots\cdots\cdots\cdots\cdots\cdots\cdots\cdots\\
\mathbf{e}'_n &= a_{1n}\mathbf{e}_1 + a_{2n}\mathbf{e}_2 + \cdots + a_{nn}\mathbf{e}_n.
\end{aligned} \tag{6}$$

The determinant of the matrix $\mathscr{A}$ in (6) is different from zero (otherwise the vectors $\mathbf{e}'_1, \mathbf{e}'_2, \cdots, \mathbf{e}'_n$ would be linearly dependent).

Let ξ_i be the coordinates of a vector $\mathbf{x}$ in the first basis and ξ'_i its coordinates in the second basis. Then

$$\mathbf{x} = \xi_1\mathbf{e}_1 + \xi_2\mathbf{e}_2 + \cdots + \xi_n\mathbf{e}_n = \xi'_1\mathbf{e}'_1 + \xi'_2\mathbf{e}'_2 + \cdots + \xi'_n\mathbf{e}'_n.$$

Replacing the $\mathbf{e}'_i$ with the appropriate expressions from (6) we get

$$\begin{aligned}
\mathbf{x} = \xi_1\mathbf{e}_1 + \xi_2\mathbf{e}_2 + \cdots + \xi_n\mathbf{e}_n &= \xi'_1(a_{11}\mathbf{e}_1 + a_{21}\mathbf{e}_2 + \cdots + a_{n1}\mathbf{e}_n)\\
&+ \xi'_2(a_{12}\mathbf{e}_1 + a_{22}\mathbf{e}_2 + \cdots + a_{n2}\mathbf{e}_n)\\
&+ \cdots\cdots\cdots\cdots\cdots\cdots\\
&+ \xi'_n(a_{1n}\mathbf{e}_1 + a_{2n}\mathbf{e}_2 + \cdots + a_{nn}\mathbf{e}_n).
\end{aligned}$$

Since the $\mathbf{e}_i$ are linearly independent, the coefficients of the $\mathbf{e}_i$ on both sides of the above equation must be the same. Hence

$$\begin{aligned} \xi_1 &= a_{11}\xi'_1 + a_{12}\xi'_2 + \cdots + a_{1n}\xi'_n, \\ \xi_2 &= a_{21}\xi'_1 + a_{22}\xi'_2 + \cdots + a_{2n}\xi'_n, \\ &\cdots\cdots\cdots\cdots\cdots\cdots\cdots\cdots \\ \xi_n &= a_{n1}\xi'_1 + a_{n2}\xi'_2 + \cdots + a_{nn}\xi'_n. \end{aligned} \tag{7}$$

Thus the coordinates ξ_i of the vector $\mathbf{x}$ in the first basis are expressed through its coordinates in the second basis by means of the matrix $\mathscr{A}'$ which is the transpose of $\mathscr{A}$.

To rephrase our result we solve the system (7) for $\xi'_1, \xi'_2, \cdots, \xi'_n$. Then

$$\begin{aligned} \xi'_1 &= b_{11}\xi_1 + b_{12}\xi_2 + \cdots + b_{1n}\xi_n, \\ \xi'_2 &= b_{21}\xi_1 + b_{22}\xi_2 + \cdots + b_{2n}\xi_n, \\ &\cdots\cdots\cdots\cdots\cdots\cdots\cdots\cdots \\ \xi'_n &= b_{n1}\xi_1 + b_{n2}\xi_2 + \cdots + b_{nn}\xi_n, \end{aligned}$$

where the b_{ik} are the elements of the inverse of the matrix $\mathscr{A}'$. Thus, *the coordinates of a vector are transformed by means of a matrix $\mathscr{B}$ which is the inverse of the transpose of the matrix $\mathscr{A}$ in* (6) *which determines the change of basis.*

§ 2. *Euclidean space*

1. *Definition of Euclidean space.* In the preceding section a vector space was defined as a collection of elements (vectors) for which there are defined the operations of addition and multiplication by scalars.

By means of these operations it is possible to define in a vector space the concepts of line, plane, dimension, parallelism of lines, etc. However, many concepts of so-called Euclidean geometry cannot be formulated in terms of addition and multiplication by scalars. Instances of such concepts are: length of a vector, angles between vectors, the inner product of vectors. The simplest way of introducing these concepts is the following.

We take as our fundamental concept the concept of an inner product of vectors. We define this concept axiomatically. Using the inner product operation in addition to the operations of addition and multiplication by scalars we shall find it possible to develop all of Euclidean geometry.

DEFINITION 1. *If with every pair of vectors* $\mathbf{x}$, $\mathbf{y}$ *in a real vector space* $\mathbf{R}$ *there is associated a real number* $(\mathbf{x}, \mathbf{y})$ *such that*

1. $(\mathbf{x}, \mathbf{y}) = (\mathbf{y}, \mathbf{x})$,
2. $(\lambda\mathbf{x}, \mathbf{y}) = \lambda(\mathbf{x}, \mathbf{y})$, $\quad$ (λ real)
3. $(\mathbf{x}_1 + \mathbf{x}_2, \mathbf{y}) = (\mathbf{x}_1, \mathbf{y}] + (\mathbf{x}_2, \mathbf{y}]$,
4. $(\mathbf{x}, \mathbf{x}) \geqq 0$ *and* $(\mathbf{x}, \mathbf{x}) = 0$ *if and only if* $\mathbf{x} = \mathbf{0}$,

then we say that an inner product is defined in $\mathbf{R}$.

A vector space in which an inner product satisfying conditions 1 through 4 has been defined is referred to as a Euclidean space.

EXAMPLES. *1.* Let us consider the (three-dimensional) space $\mathbf{R}$ of vectors studied in elementary solid geometry (cf. Example *1*, § 1). Let us define the inner product of two vectors in this space as the product of their lengths by the cosine of the angle between them. We leave it to the reader to verify the fact that the operation just defined satisfies conditions 1 through 4 above.

2. Consider the space $\mathbf{R}$ of n-tuples of real numbers. Let $\mathbf{x} = (\xi_1, \xi_2, \cdots, \xi_n)$ and $\mathbf{y} = (\eta_1, \eta_2, \cdots, \eta_n)$ be in $\mathbf{R}$. In addition to the definitions of addition

$$\mathbf{x} + \mathbf{y} = (\xi_1 + \eta_1, \xi_2 + \eta_2, \cdots, \xi_n + \eta_n)$$

and multiplication by scalars

$$\lambda\mathbf{x} = (\lambda\xi_1, \lambda\xi_2, \cdots, \lambda\xi_n)$$

with which we are already familiar from Example *2*, § 1, we define the inner product of $\mathbf{x}$ and $\mathbf{y}$ as

$$(\mathbf{x}, \mathbf{y}) = \xi_1\eta_1 + \xi_2\eta_2 + \cdots + \xi_n\eta_n.$$

It is again easy to check that properties 1 through 4 are satisfied by $(\mathbf{x}, \mathbf{y})$ as defined.

3. Without changing the definitions of addition and multiplication by scalars in Example *2* above we shall define the inner product of two vectors in the space of Example *2* in a different and more general manner.

Thus let $||a_{ik}||$ be a real $n \times n$ matrix. Let us put

$$(1)\qquad \begin{aligned}(\mathbf{x}, \mathbf{y}) = a_{11}\xi_1\eta_1 &+ a_{12}\xi_1\eta_2 + \cdots + a_{1n}\xi_1\eta_n \\ &+ a_{21}\xi_2\eta_1 + a_{22}\xi_2\eta_2 + \cdots + a_{2n}\xi_2\eta_n \\ &\cdots\cdots\cdots\cdots\cdots\cdots\cdots\cdots \\ &+ a_{n1}\xi_n\eta_1 + a_{n2}\xi_n\eta_2 + \cdots + a_{nn}\xi_n\eta_n.\end{aligned}$$

We can verify directly the fact that this definition satisfies Axioms 2 and 3 for an inner product regardless of the nature of the real matrix $||a_{ik}||$. For Axiom 1 to hold, that is, for $(\mathbf{x}, \mathbf{y})$ to be symmetric relative to $\mathbf{x}$ and $\mathbf{y}$, it is necessary and sufficient that

$$(2)\qquad a_{ik} = a_{ki},$$

i.e., that $||a_{ik}||$ be symmetric.

Axiom 4 requires that the expression

$$(3)\qquad (\mathbf{x}, \mathbf{x}) = \sum_{i,k=1}^{n} a_{ik}\xi_i\xi_k$$

be non-negative fore very choice of the n numbers $\xi_1, \xi_2, \cdots, \xi_n$ and that it vanish only if $\xi_1 = \xi_2 = \cdots = \xi_n = 0$.

The homogeneous polynomial or, as it is frequently called, quadratic form in (3) is said to be *positive definite* if it takes on non-negative values only and if it vanishes only when all the ξ_i are zero. Thus for Axiom 4 to hold the quadratic form (3) must be positive definite.

In summary, for (1) to define an inner product the matrix $||a_{ik}||$ must be symmetric and the quadratic form associated with $||a_{ik}||$ must be positive definite.

If we take as the matrix $||a_{ik}||$ the unit matrix, i.e., if we put $a_{ii} = 1$ and $a_{ik} = 0$ $(i \neq k)$, then the inner product $(\mathbf{x}, \mathbf{y})$ defined by (1) takes the form

$$(\mathbf{x}, \mathbf{y}) = \sum_{i=1}^{n} \xi_i\eta_i$$

and the result is the Euclidean space of Example 2.

EXERCISE. Show that the matrix

$$\begin{pmatrix} 0 & 1 \\ 1 & 0 \end{pmatrix}$$

cannot be used to define an inner product (the corresponding quadratic form is not positive definite), and that the matrix

$$\begin{pmatrix} 1 & 1 \\ 1 & 2 \end{pmatrix}$$

can be used to define an inner product satisfying the axioms 1 through 4.

In the sequel (§ 6) we shall give simple criteria for a quadratic form to be positive definite.

4. Let the elements of a vector space be all the continuous functions on an interval $[a, b]$. We define the inner product of two such functions as the integral of their product

$$(f, g) = \int_a^b f(t)g(t)\, dt.$$

It is easy to check that the Axioms 1 through 4 are satisfied.

5. Let **R** be the space of polynomials of degree $\leq n - 1$. We define the inner product of two polynomials as in Example *4*

$$(P, Q) = \int_a^b P(t)Q(t)\, dt.$$

2. *Length of a vector. Angle between two vectors.* We shall now make use of the concept of an inner product to define the length of a vector and the angle between two vectors.

DEFINITION 2. *By the length of a vector* **x** *in Euclidean space we mean the number*

$$\sqrt{(\mathbf{x}, \mathbf{x})}. \tag{4}$$

We shall denote the length of a vector **x** by the symbol $|\mathbf{x}|$.

It is quite natural to require that the definitions of length of a vector, of the angle between two vectors and of the inner product of two vectors imply the usual relation which connects these quantities. In other words, it is natural to require that the inner product of two vectors be equal to the product of the lengths of these vectors times the cosine of the angle between them. This dictates the following definition of the concept of angle between two vectors.

DEFINITION 3. *By the angle between two vectors* **x** *and* **y** *we mean the number*

$$\varphi = \arc\cos \frac{(\mathbf{x}, \mathbf{y})}{|\mathbf{x}|\,|\mathbf{y}|};$$

i.e., we put

$$\cos \varphi = \frac{(\mathbf{x}, \mathbf{y})}{|\mathbf{x}|\,|\mathbf{y}|}. \tag{5}$$

The vectors $\mathbf{x}$ and $\mathbf{y}$ are said to be *orthogonal* if $(\mathbf{x}, \mathbf{y}) = 0$. The angle between two non-zero orthogonal vectors is clearly $\pi/2$. The concepts just introduced permit us to extend a number of theorems of elementary geometry to Euclidean spaces. [1]

The following is an example of such extension. If $\mathbf{x}$ and $\mathbf{y}$ are orthogonal vectors, then it is natural to regard $\mathbf{x} + \mathbf{y}$ as the diagonal of a rectangle with sides $\mathbf{x}$ and $\mathbf{y}$. We shall show that

$$|\mathbf{x} + \mathbf{y}|^2 = |\mathbf{x}|^2 + |\mathbf{y}|^2,$$

i.e., that *the square of the length of the diagonal of a rectangle is equal to the sum of the squares of the lengths of its two non-parallel sides* (the theorem of Pythagoras).

Proof: By definition of length of a vector

$$|\mathbf{x} + \mathbf{y}|^2 = (\mathbf{x} + \mathbf{y}, \mathbf{x} + \mathbf{y}).$$

In view of the distributivity property of inner products (Axiom 3),

$$(\mathbf{x} + \mathbf{y}, \mathbf{x} + \mathbf{y}) = (\mathbf{x}, \mathbf{x}) + (\mathbf{x}, \mathbf{y}) + (\mathbf{y}, \mathbf{x}) + (\mathbf{y}, \mathbf{y}).$$

Since $\mathbf{x}$ and $\mathbf{y}$ are supposed orthogonal,

$$(\mathbf{x}, \mathbf{y}) = (\mathbf{y}, \mathbf{x}) = 0.$$

Thus

$$|\mathbf{x} + \mathbf{y}|^2 = (\mathbf{x}, \mathbf{x}) + (\mathbf{y}, \mathbf{y}) = |\mathbf{x}|^2 + |\mathbf{y}|^2,$$

which is what we set out to prove.

This theorem can be easily generalized to read: *if* $\mathbf{x}, \mathbf{y}, \mathbf{z}, \cdots$ *are pairwise orthogonal, then*

$$|\mathbf{x} + \mathbf{y} + \mathbf{z} + \cdots|^2 = |\mathbf{x}|^2 + |\mathbf{y}|^2 + |\mathbf{z}|^2 + \cdots.$$

3. *The Schwarz inequality.* In para. 2. we defined the angle φ between two vectors $\mathbf{x}$ and $\mathbf{y}$ by means of the relation

$$\cos\varphi = \frac{(\mathbf{x}, \mathbf{y})}{|\mathbf{x}|\,|\mathbf{y}|}.$$

If φ is to be always computable from this relation we must show that

[1] We could have axiomatized the notions of length of a vector and angle between two vectors rather than the notion of inner product. However, this course would have resulted in a more complicated system of axioms than that associated with the notion of an inner product.

$$-1 \leqq \frac{(\mathbf{x}, \mathbf{y})}{|\mathbf{x}|\,|\mathbf{y}|} \leqq 1,$$

or, equivalently, that

$$\frac{(\mathbf{x}, \mathbf{y})^2}{|\mathbf{x}|^2\,|\mathbf{y}|^2} \leqq 1,$$

which, in turn, is the same as

$$(\mathbf{x}, \mathbf{y})^2 \leqq (\mathbf{x}, \mathbf{x})(\mathbf{y}, \mathbf{y}). \tag{6}$$

Inequality (6) is known as the *Schwarz inequality.*

Thus, *before we can correctly define the angle between two vectors by means of the relation* (5) *we must prove the Schwarz inequality.* [2]

To prove the Schwarz inequality we consider the vector $\mathbf{x} - t\mathbf{y}$ where t is any real number. In view of Axiom 4 for inner products,

$$(\mathbf{x} - t\mathbf{y}, \mathbf{x} - t\mathbf{y}) \geqq 0;$$

i.e., for any t,

$$t^2(\mathbf{y}, \mathbf{y}) - 2t(\mathbf{x}, \mathbf{y}) + (\mathbf{x}, \mathbf{x}) \geqq 0.$$

This inequality implies that the polynomial cannot have two distinct real roots. Consequently, the discriminant of the equation

$$t^2(\mathbf{y}, \mathbf{y}) - 2t(\mathbf{x}, \mathbf{y}) + (\mathbf{x}, \mathbf{x}) = 0$$

cannot be positive; i.e.,

$$(\mathbf{x}, \mathbf{y})^2 - (\mathbf{x}, \mathbf{x})(\mathbf{y}, \mathbf{y}) \leqq 0,$$

which is what we wished to prove.

EXERCISE. Prove that a necessary and sufficient condition for $(\mathbf{x}, \mathbf{y})^2 = (\mathbf{x}, \mathbf{x})(\mathbf{y}, \mathbf{y})$ is the linear dependence of the vectors $\mathbf{x}$ and $\mathbf{y}$.

EXAMPLES. We have proved the validity of (6) for an axiomatically defined Euclidean space. It is now appropriate to interpret this inequality in the various concrete Euclidean spaces in para. 1.

1. In the case of Example *1*, inequality (6) tells us nothing new. (cf. the remark preceding the proof of the Schwarz inequality.)

[2] Note, however that in para. 1, Example *1*, of this section there is no need to prove this inequality. Namely, in vector analysis the inner product of two vectors is defined in such a way that the quantity $(\mathbf{x}, \mathbf{y})/|\mathbf{x}|\,|\mathbf{y}|$ is the cosine of a previously determined angle between the vectors. Consequently, $|(\mathbf{x}, \mathbf{y})|/|\mathbf{x}|\,|\mathbf{y}| \leqq 1$.

2. In Example *2* the inner product was defined as

$$(\mathbf{x}, \mathbf{y}) = \sum_{i=1}^{n} \xi_i \eta_i .$$

It follows that

$$(\mathbf{x}, \mathbf{x}) = \sum_{i=1}^{n} \xi_i^2, \quad (\mathbf{y}, \mathbf{y}) = \sum_{i=1}^{n} \eta_i^2,$$

and inequality (6) becomes

$$\left(\sum_{i=1}^{n} \xi_i \eta_i\right)^2 \leqq \left(\sum_{i=1}^{n} \xi_i^2\right)\left(\sum_{i=1}^{n} \eta_i^2\right).$$

3. In Example *3* the inner product was defined as

$$(1) \qquad (\mathbf{x}, \mathbf{y}) = \sum_{i,k=1}^{n} a_{ik} \xi_i \eta_k,$$

where

$$(2) \qquad a_{ik} = a_{ki}$$

and

$$(3) \qquad \sum_{i,k=1}^{n} a_{ik} \xi_i \xi_k \geqq 0$$

for any choice of the ξ_i. Hence (6) implies that

If the numbers a_{ik} satisfy conditions (2) *and* (3), *then the following inequality holds*:

$$\left(\sum_{i,k=1}^{n} a_{ik} \xi_i \eta_k\right)^2 \leqq \left(\sum_{i,k=1}^{n} a_{ik} \xi_i \xi_k\right)\left(\sum_{i,k=1}^{n} a_{ik} \eta_i \eta_k\right).$$

EXERCISE. Show that if the numbers a_{ik} satisfy conditions (2) and (3), $a_{ik}^2 \leqq a_{ii} a_{kk}$. (*Hint*: Assign suitable values to the numbers $\xi_1, \xi_2, \cdots, \xi_n$ and $\eta_1, \eta_2, \cdots, \eta_n$ in the inequality just derived.)

4. In Example *4* the inner product was defined by means of the integral $\int_a^b f(t) g(t)\, dt$. Hence (6) takes the form

$$\left(\int_a^b f(t) g(t)\, dt\right)^2 \leqq \int_a^b [f(t)]^2\, dt \cdot \int_a^b [g(t)]^2\, dt.$$

This inequality plays an important role in many problems of analysis.

We now give an example of an inequality which is a consequence of the Schwarz inequality.

If $\mathbf{x}$ *and* $\mathbf{y}$ *are two vectors in a Euclidean space* $\mathbf{R}$ *then*

$$(7) \qquad |\mathbf{x} + \mathbf{y}| \leqq |\mathbf{x}| + |\mathbf{y}|.$$

Proof:

$$|\mathbf{x} + \mathbf{y}|^2 = (\mathbf{x} + \mathbf{y}, \mathbf{x} + \mathbf{y}) = (\mathbf{x}, \mathbf{x}) + 2(\mathbf{x}, \mathbf{y}) + (\mathbf{y}, \mathbf{y}).$$

Since $2(\mathbf{x}, \mathbf{y}) \leqq 2|\mathbf{x}|\,|\mathbf{y}|$, it follows that

$$|\mathbf{x}+\mathbf{y}|^2 = (\mathbf{x}+\mathbf{y}, \mathbf{x}+\mathbf{y}) \leqq (\mathbf{x}, \mathbf{x})+2|\mathbf{x}|\,|\mathbf{y}|+(\mathbf{y}, \mathbf{y}) = (|\mathbf{x}|+|\mathbf{y}|)^2,$$

i.e., $|\mathbf{x} + \mathbf{y}| \leqq |\mathbf{x}| + |\mathbf{y}|$, which is the desired conclusion.

EXERCISE. Interpret inequality (7) in each of the concrete Euclidean spaces considered in the beginning of this section.

In geometry the distance between two points $\mathbf{x}$ and $\mathbf{y}$ (note the use of the same symbol to denote a vector—drawn from the origin—and a point, the tip of that vector) is defined as the length of the vector $\mathbf{x} - \mathbf{y}$. In the general case of an n-dimensional Euclidean space *we define the distance between* $\mathbf{x}$ *and* $\mathbf{y}$ *by the relation*

$$d = |\mathbf{x} - \mathbf{y}|.$$

§ 3. Orthogonal basis. Isomorphism of Euclidean spaces

1. *Orthogonal basis.* In § 1 we introduced the notion of a basis (coordinate system) of a vector space. In a vector space there is no reason to prefer one basis to another. [3] Not so in Euclidean spaces. Here there is every reason to prefer so-called orthogonal bases to all other bases. Orthogonal bases play the same role in Euclidean spaces which rectangular coordinate systems play in analytic geometry.

DEFINITION 1. *The non-zero vectors* $\mathbf{e}_1, \mathbf{e}_2, \cdots, \mathbf{e}_n$ *of an n-dimensional Euclidean vector space are said to form an orthogonal basis if they are pairwise orthogonal, and an orthonormal basis if, in addition, each has unit length. Briefly, the vectors* $\mathbf{e}_1, \mathbf{e}_2, \cdots, \mathbf{e}_n$ *form an orthonormal basis if*

[3] Careful reading of the proof of the isomorphism of vector spaces given in § 1 will show that in addition to proving the theorem we also showed that it is possible to construct an isomorphism of two n-dimensional vector spaces which takes a specified basis in one of these spaces into a specified basis in the other space. In particular, if $\mathbf{e}_1, \mathbf{e}_2, \cdots, \mathbf{e}_n$ and $\mathbf{e}'_1, \mathbf{e}'_2, \cdots, \mathbf{e}'_n$ are two bases in $\mathbf{R}$, then there exists an isomorphic mapping of $\mathbf{R}$ onto itself which takes the first of these bases into the second.

(1) $$(\mathbf{e}_i, \mathbf{e}_k) = \begin{cases} 1 & \text{if} \quad i = k \\ 0 & \text{if} \quad i \neq k. \end{cases}$$

For this definition to be correct we must prove that the vectors $\mathbf{e}_1, \mathbf{e}_2, \cdots, \mathbf{e}_n$ of the definition actually form a basis, i.e., are linearly independent.

Thus, let

(2) $$\lambda_1 \mathbf{e}_1 + \lambda_2 \mathbf{e}_2 + \cdots + \lambda_n \mathbf{e}_n = \mathbf{0}.$$

We wish to show that (2) implies $\lambda_1 = \lambda_2 = \cdots = \lambda_n = 0$. To this end we multiply both sides of (2) by $\mathbf{e}_1$ (i.e., form the inner product of each side of (2) with $\mathbf{e}_1$). The result is

$$\lambda_1(\mathbf{e}_1, \mathbf{e}_1) + \lambda_2(\mathbf{e}_1, \mathbf{e}_2) + \cdots + \lambda_n(\mathbf{e}_1, \mathbf{e}_n) = 0.$$

Now, the definition of an orthogonal basis implies that

$$(\mathbf{e}_1, \mathbf{e}_1) \neq 0, \qquad (\mathbf{e}_1, \mathbf{e}_k) = 0 \qquad \text{for } k \neq 1.$$

Hence $\lambda_1 = 0$. Likewise, multiplying (2) by $\mathbf{e}_2$ we find that $\lambda_2 = 0$, etc. This proves that $\mathbf{e}_1, \mathbf{e}_2, \cdots, \mathbf{e}_n$ are linearly independent.

We shall make use of the so-called *orthogonalization procedure* to prove the existence of orthogonal bases. This procedure leads from any basis $\mathbf{f}_1, \mathbf{f}_2, \cdots, \mathbf{f}_n$ to an orthogonal basis $\mathbf{e}_1, \mathbf{e}_2, \cdots, \mathbf{e}_n$.

THEOREM 1. *Every n-dimensional Euclidean space contains orthogonal bases.*

Proof: By definition of an n-dimensional vector space (§ 1, para. 2) such a space contains a basis $\mathbf{f}_1, \mathbf{f}_2, \cdots, \mathbf{f}_n$. We put $\mathbf{e}_1 = \mathbf{f}_1$. Next we put $\mathbf{e}_2 = \mathbf{f}_2 + \alpha \mathbf{e}_1$, where α is chosen so that $(\mathbf{e}_2, \mathbf{e}_1) = 0$; i.e., $(\mathbf{f}_2 + \alpha \mathbf{e}_1, \mathbf{e}_1) = 0$. This means that

$$\alpha = -(\mathbf{f}_2, \mathbf{e}_1)/(\mathbf{e}_1, \mathbf{e}_1).$$

Suppose that we have already constructed non-zero pairwise orthogonal vectors $\mathbf{e}_1, \mathbf{e}_2, \cdots, \mathbf{e}_{k-1}$. To construct $\mathbf{e}_k$ we put

(3) $$\mathbf{e}_k = \mathbf{f}_k + \lambda_1 \mathbf{e}_{k-1} + \cdots + \lambda_{k-1} \mathbf{e}_1,$$

where the λ_k are determined from the orthogonality conditions

$$\begin{aligned}
(\mathbf{e}_k, \mathbf{e}_1) &= (\mathbf{f}_k + \lambda_1 \mathbf{e}_{k-1} + \cdots + \lambda_{k-1} \mathbf{e}_1, \mathbf{e}_1) &= 0,\\
(\mathbf{e}_k, \mathbf{e}_2) &= (\mathbf{f}_k + \lambda_1 \mathbf{e}_{k-1} + \cdots + \lambda_{k-1} \mathbf{e}_1, \mathbf{e}_2) &= 0,\\
&\cdots\cdots\cdots\cdots\cdots\cdots\cdots\cdots\\
(\mathbf{e}_k, \mathbf{e}_{k-1}) &= (\mathbf{f}_k + \lambda_1 \mathbf{e}_{k-1} + \cdots + \lambda_{k-1} \mathbf{e}_1, \mathbf{e}_{k-1}) &= 0.
\end{aligned}$$

Since the vectors $\mathbf{e}_1, \mathbf{e}_2, \cdots, \mathbf{e}_{k-1}$ are pairwise orthogonal, the latter equalities become:

$$\begin{aligned}
(\mathbf{f}_k, \mathbf{e}_1) + \lambda_{k-1}(\mathbf{e}_1, \mathbf{e}_1) &= 0,\\
(\mathbf{f}_k, \mathbf{e}_2) + \lambda_{k-2}(\mathbf{e}_2, \mathbf{e}_2) &= 0,\\
&\cdots\cdots\cdots\\
(\mathbf{f}_k, \mathbf{e}_{k-1}) + \lambda_1(\mathbf{e}_{k-1}, \mathbf{e}_{k-1}) &= 0.
\end{aligned}$$

It follows that

$$(4) \quad \lambda_{k-1} = -(\mathbf{f}_k, \mathbf{e}_1)/(\mathbf{e}_1, \mathbf{e}_1), \lambda_{k-2} = -(\mathbf{f}_k, \mathbf{e}_2)/(\mathbf{e}_2, \mathbf{e}_2), \cdots, \\ \lambda_1 = -(\mathbf{f}_k, \mathbf{e}_{k-1})/(\mathbf{e}_{k-1}, \mathbf{e}_{k-1}).$$

So far we have not made use of the linear independence of the vectors $\mathbf{f}_1, \mathbf{f}_2, \cdots, \mathbf{f}_n$, but we shall make use of this fact presently to prove that $\mathbf{e}_k \neq 0$. The vector $\mathbf{e}_k$ is a linear combination of the vectors $\mathbf{e}_1, \mathbf{e}_2, \cdots, \mathbf{e}_{k-1}, \mathbf{f}_k$. But $\mathbf{e}_{k-1}$ can be written as a linear combination of the vector $\mathbf{f}_{k-1}$ and the vectors $\mathbf{e}_1, \mathbf{e}_2, \cdots, \mathbf{e}_{k-2}$. Similar statements hold for $\mathbf{e}_{k-2}, \mathbf{e}_{k-3}, \cdots, \mathbf{e}_1$. It follows that

$$(5) \qquad \mathbf{e}_k = a_1 \mathbf{f}_1 + a_2 \mathbf{f}_2 + \cdots + a_{k-1} \mathbf{f}_{k-1} + \mathbf{f}_k.$$

In view of the linear independence of the vectors $\mathbf{f}_1, \mathbf{f}_2, \cdots, \mathbf{f}_k$ we may conclude on the basis of eq. (5) that $\mathbf{e}_k \neq 0$.

Just as $\mathbf{e}_1, \mathbf{e}_2, \cdots, \mathbf{e}_{k-1}$ and $\mathbf{f}_k$ were used to construct $\mathbf{e}_k$ so $\mathbf{e}_1, \mathbf{e}_2, \cdots, \mathbf{e}_k$ and $\mathbf{f}_{k+1}$ can be used to construct $\mathbf{e}_{k+1}$, etc.

By continuing the process described above we obtain n non-zero, pairwise orthogonal vectors $\mathbf{e}_1, \mathbf{e}_2, \cdots, \mathbf{e}_n$, i.e., an orthogonal basis. This proves our theorem.

It is clear that the vectors

$$\mathbf{e}'_k = \mathbf{e}_k/|\mathbf{e}_k| \qquad (k = 1, 2, \cdots, n)$$

form an orthonormal basis.

Examples of orthogonalization. *1.* Let $\mathbf{R}$ be the three-dimensional space with which we are familiar from elementary geometry. Let $\mathbf{f}_1, \mathbf{f}_2, \mathbf{f}_3$ be three linearly independent vectors in $\mathbf{R}$. Put $\mathbf{e}_1 = \mathbf{f}_1$. Next select a vector $\mathbf{e}_2$ perpendicular to $\mathbf{e}_1$ and lying in the plane determined by $\mathbf{e}_1 = \mathbf{f}_1$

and $\mathbf{f}_2$. Finally, choose $\mathbf{e}_3$ perpendicular to $\mathbf{e}_1$ and $\mathbf{e}_2$ (i.e., perpendicular to the previously constructed plane).

2. Let **R** be the three-dimensional vector space of polynomials of degree not exceeding two. We define the inner product of two vectors in this space by the integral

$$\int_{-1}^{1} P(t)Q(t)\,dt.$$

The vectors 1, t, t^2 form a basis in **R**. We shall now orthogonalize this basis. We put $\mathbf{e}_1 = 1$. Next we put $\mathbf{e}_2 = t + \alpha \cdot 1$. Since

$$0 = (t+\alpha \cdot 1, 1) = \int_{-1}^{1} (t + \alpha)\,dt = 2\alpha,$$

it follows that $\alpha = 0$, i.e., $\mathbf{e}_2 = t$. Finally we put $\mathbf{e}_3 = t^2 + \beta t + \gamma \cdot 1$. The orthogonality requirements imply $\beta = 0$ and $\gamma = -1/3$, i.e., $\mathbf{e}_3 = t^2 - 1/3$. Thus 1, t, $t^2 - 1/3$ is an orthogonal basis in **R**. By dividing each basis vector by its length we obtain an orthonormal basis for **R**.

3. Let **R** be the space of polynomials of degree not exceeding $n - 1$. We define the inner product of two vectors in this space as in the preceding example.

We select as basis the vectors 1, $t, \cdots, t^{n-1}$. As in Example *2* the process of orthogonalization leads to the sequence of polynomials

$$1,\ t,\ t^2 - 1/3,\ t^3 - (3/5)t,\ \cdots.$$

Apart from multiplicative constants these polynomials coincide with the Legendre polynomials

$$\frac{1}{2^k \cdot k!} \frac{d^k (t^2 - 1)^k}{dt^k}.$$

The Legendre polynomials form an orthogonal, but not orthonormal basis in **R**. Multiplying each Legendre polynomial by a suitable constant we obtain an orthonormal basis in **R**. We shall denote the kth element of this basis by $P_k(t)$.

Let $\mathbf{e}_1, \mathbf{e}_2, \cdots, \mathbf{e}_n$ be an orthonormal basis of a Euclidean space **R**. If

$$\mathbf{x} = \xi_1\mathbf{e}_1 + \xi_2\mathbf{e}_2 + \cdots + \xi_n\mathbf{e}_n,$$
$$\mathbf{y} = \eta_1\mathbf{e}_1 + \eta_2\mathbf{e}_2 + \cdots + \eta_n\mathbf{e}_n,$$

then

$$(\mathbf{x}, \mathbf{y}) = (\xi_1\mathbf{e}_1 + \xi_2\mathbf{e}_2 + \cdots + \xi_n\mathbf{e}_n, \eta_1\mathbf{e}_1 + \eta_2\mathbf{e}_2 + \cdots + \eta_n\mathbf{e}_n).$$

Since

$$(\mathbf{e}_i, \mathbf{e}_k) = \begin{cases} 1 & \text{if} \quad i = k \\ 0 & \text{if} \quad i \neq k, \end{cases}$$

it follows that

$$(\mathbf{x}, \mathbf{y}) = \xi_1\eta_1 + \xi_2\eta_2 + \cdots + \xi_n\eta_n.$$

Thus, *the inner product of two vectors relative to an orthonormal basis is equal to the sum of the products of the corresponding coordinates of these vectors* (cf. Example 2, § 2).

EXERCISES. *1.* Show that if $\mathbf{f}_1, \mathbf{f}_2, \cdots, \mathbf{f}_n$ is an arbitrary basis, then

$$(\mathbf{x}, \mathbf{y}) = \sum_{i,k=1}^{n} a_{ik}\xi_i\eta_k,$$

where $a_{ik} = a_{ki}$ and $\xi_1, \xi_2, \cdots, \xi_n$ and $\eta_1, \eta_2, \cdots, \eta_n$ are the coordinates of $\mathbf{x}$ and $\mathbf{y}$ respectively.

2. Show that if in some basis $\mathbf{f}_1, \mathbf{f}_2, \cdots, \mathbf{f}_n$

$$(\mathbf{x}, \mathbf{y}) = \xi_1\eta_1 + \xi_2\eta_2 + \cdots + \xi_n\eta_n$$

for every $\mathbf{x} = \xi_1\mathbf{f}_1 + \cdots + \xi_n\mathbf{f}_n$ and $\mathbf{y} = \eta_1\mathbf{f}_1 + \cdots + \eta_n\mathbf{f}_n$, then this basis is orthonormal.

We shall now find the coordinates of a vector $\mathbf{x}$ relative to an orthonormal basis $\mathbf{e}_1, \mathbf{e}_2, \cdots, \mathbf{e}_n$.

Let

$$\mathbf{x} = \xi_1\mathbf{e}_1 + \xi_2\mathbf{e}_2 + \cdots + \xi_n\mathbf{e}_n.$$

Multiplying both sides of this equation by $\mathbf{e}_1$ we get

$$(\mathbf{x}, \mathbf{e}_1) = \xi_1(\mathbf{e}_1, \mathbf{e}_1) + \xi_2(\mathbf{e}_2\, \mathbf{e}_1) + \cdots + \xi_n(\mathbf{e}_n, \mathbf{e}_1) = \xi_1$$

and, similarly,

$$\xi_2 = (\mathbf{x}, \mathbf{e}_2), \cdots, \xi_n = (\mathbf{x}, \mathbf{e}_n). \tag{7}$$

Thus *the kth coordinate of a vector relative to an orthonormal basis is the inner product of this vector and the kth basis vector.*

It is natural to call the inner product of a vector $\mathbf{x}$ and a vector $\mathbf{e}$ of length 1 *the projection* of $\mathbf{x}$ on $\mathbf{e}$. The result just proved may be states as follows: *The coordinates of a vector relative to an orthonormal basis are the projections of this vector on the basis vectors.* This is the exact analog of a statement with which we are familiar from analytic geometry, except that there we speak of projections on the coordinate axes rather than on the basis vectors.

EXAMPLES. *1.* Let $P_0(t), P_1(t), \cdots, P_n(t)$ be the normed Legendre polynomials of degree $0, 1, \cdots, n$. Further, let $Q(t)$ be an arbitrary polynomial of degree n. We shall represent $Q(t)$ as a linear combination of the Legendre polynomials. To this end we note that all polynomials of degree $\leqq n$ form an n-dimensional vector space with orthonormal basis $P_0(t)$, $P_1(t), \cdots, P_n(t)$. Hence every polynomial $Q(t)$ of degree $\leqq n$ can be represented in the form

$$Q(t) = c_0 P_0(t) + c_1 P_1(t) + \cdots + c_n P_n(t).$$

It follows from (7) that

$$c_i = \int_{-1}^{1} Q(t) P_i(t)\, dt.$$

2. Consider the system of functions

$$(8) \qquad 1, \cos t, \sin t, \cos 2t, \sin 2t, \cdots, \cos nt, \sin nt,$$

on the interval $(0, 2\pi)$. A linear combination

$$P(t) = (a_0/2) + a_1 \cos t + b_1 \sin t + a_2 \cos 2t + \cdots + b_n \sin nt$$

of these functions is called a trigonometric polynomial of degree n. The totality of trigonometric polynomials of degree n form a $(2n + 1)$-dimensional space $\mathbf{R}_1$. We define an inner product in $\mathbf{R}_1$ by the usual integral

$$(P, Q) = \int_0^{2\pi} P(t) Q(t)\, dt.$$

It is easy to see that the system (8) is an orthogonal basis. Indeed

$$\int_0^{2\pi} \cos kt \cos lt\, dt = 0 \qquad \text{if } k \neq l,$$

$$\int_0^{2\pi} \sin kt \cos lt\, dt = 0,$$

$$\int_0^{2\pi} \sin kt \sin lt\, dt = 0, \qquad \text{if } k \neq l.$$

Since

$$\int_0^{2\pi} \sin^2 kt\, dt = \int_0^{2\pi} \cos^2 kt\, dt = \pi \qquad \text{and} \qquad \int_0^{2\pi} 1 dt = 2\pi,$$

it follows that the functions

$$(8') \quad 1/\sqrt{2\pi},\ (1/\sqrt{\pi}) \cos t,\ (1/\sqrt{\pi}) \sin t, \cdots, (1/\sqrt{\pi}) \cos nt,\ (1/\sqrt{\pi}) \sin nt$$

are an orthonormal basis for $\mathbf{R}_1$.

2. *Perpendicular from a point to a subspace. The shortest distance from a point to a subspace.* (This paragraph may be left out in a first reading.)

DEFINITION 2. *Let $\mathbf{R}_1$ be a subspace of a Euclidean space $\mathbf{R}$. We shall say that a vector $\mathbf{h} \in \mathbf{R}$ is orthogonal to the subspace $\mathbf{R}_1$ if it is orthogonal to every vector $\mathbf{x} \in \mathbf{R}_1$.*

If $\mathbf{h}$ is orthogonal to the vectors $\mathbf{e}_1, \mathbf{e}_2, \cdots, \mathbf{e}_m$, then it is also orthogonal to any linear combination of these vectors. Indeed,

$$(\mathbf{h}, \mathbf{e}_i) = 0 \quad (i = 1, 2, \cdots, m)$$

implies that for any numbers $\lambda_1, \lambda_2, \cdots, \lambda_m$

$$(\mathbf{h}, \lambda_1\mathbf{e}_1 + \lambda_2\mathbf{e}_2 + \cdots + \lambda_m\mathbf{e}_m) = 0.$$

Hence, for a vector $\mathbf{h}$ to be orthogonal to an m-dimensional subspace of $\mathbf{R}$ it is sufficient that it be orthogonal to m linearly independent vectors in $\mathbf{R}_1$, i.e., to a basis of $\mathbf{R}_1$.

Let $\mathbf{R}_1$ be an m-dimensional subspace of a (finite or infinite dimensional) Euclidean space $\mathbf{R}$ and let $\mathbf{f}$ be a vector not belonging to $\mathbf{R}_1$. We pose the problem of dropping a perpendicular from the point $\mathbf{f}$ to $\mathbf{R}_1$, i.e., of *finding a vector* $\mathbf{f}_0$ *in* $\mathbf{R}_1$ *such that the vector* $\mathbf{h} = \mathbf{f} - \mathbf{f}_0$ *is orthogonal to* $\mathbf{R}_1$. The vector $\mathbf{f}_0$ is called *the orthogonal projection of* $\mathbf{f}$ *on the subspace* $\mathbf{R}_1$. We shall see in the sequel that this problem has always a unique solution. Right now we shall show that, just as in Euclidean geometry, $|\mathbf{h}|$ is the shortest distance from $\mathbf{f}$ to $\mathbf{R}_1$. In other words, we shall show that *if* $\mathbf{f}_1 \in \mathbf{R}_1$ *and* $\mathbf{f}_1 \neq \mathbf{f}_0$, *then.*

$$|\mathbf{f} - \mathbf{f}_1| > |\mathbf{f} - \mathbf{f}_0|.$$

Indeed, as a difference of two vectors in $\mathbf{R}_1$, the vector $\mathbf{f}_0 - \mathbf{f}_1$ belongs to $\mathbf{R}_1$ and is therefore orthogonal to $\mathbf{h} = \mathbf{f} - \mathbf{f}_0$. By the theorem of Pythagoras

$$|\mathbf{f} - \mathbf{f}_0|^2 + |\mathbf{f}_0 - \mathbf{f}_1|^2 = |\mathbf{f} - \mathbf{f}_0 + \mathbf{f}_0 - \mathbf{f}_1|^2 = |\mathbf{f} - \mathbf{f}_1|^2,$$

so that

$$|\mathbf{f} - \mathbf{f}_1| > |\mathbf{f} - \mathbf{f}_0|.$$

We shall now show how one can actually compute the orthogonal projection $\mathbf{f}_0$ of $\mathbf{f}$ on the subspace $\mathbf{R}_1$ (i.e., how to drop a perpendicular from $\mathbf{f}$ on $\mathbf{R}_1$). Let $\mathbf{e}_1, \mathbf{e}_2, \cdots, \mathbf{e}_m$ be a basis of $\mathbf{R}_1$. As a vector in $\mathbf{R}_1$, $\mathbf{f}_0$ must be of the form

$$\mathbf{f}_0 = c_1\mathbf{e}_1 + c_2\mathbf{e}_2 + \cdots + c_m\mathbf{e}_m. \tag{9}$$

To find the c_k we note that $\mathbf{f} - \mathbf{f}_0$ must be orthogonal to $\mathbf{R}_1$, i.e., $(\mathbf{f} - \mathbf{f}_0, \mathbf{e}_k) = 0$ $(k = 1, 2, \cdots, m)$, or,

$$(\mathbf{f}_0, \mathbf{e}_k) = (\mathbf{f}, \mathbf{e}_k). \tag{10}$$

Replacing $\mathbf{f}_0$ by the expression in (9) we obtain a system of m equations for the c_k

$$(11)\quad c_1(\mathbf{e}_1, \mathbf{e}_k) + c_2(\mathbf{e}_2, \mathbf{e}_k) + \cdots + c_m(\mathbf{e}_m, \mathbf{e}_k) = (\mathbf{f}, \mathbf{e}_k) \qquad (k = 1, 2, \cdots, m).$$

We first consider the frequent case when the vectors $\mathbf{e}_1, \mathbf{e}_2, \cdots, \mathbf{e}_m$ are orthonormal. In this case the problem can be solved with ease. Indeed, in such a basis the system (11) goes over into the system

$$(12)\qquad c_i = (\mathbf{f}, \mathbf{e}_i).$$

Since it is always possible to select an orthonormal basis in an m-dimensional subspace, we have proved that *for every vector* $\mathbf{f}$ *there exists a unique orthogonal projection* $\mathbf{f}_0$ *on the subspace* $\mathbf{R}_1$.

We shall now show that for an arbitrary basis $\mathbf{e}_1, \mathbf{e}_2, \cdots, \mathbf{e}_m$ the system (11) must also have a unique solution. Indeed, in view of the established existence and uniqueness of the vector $\mathbf{f}_0$, this vector has uniquely determined coordinates $c_1, c_2, \cdots, c_m$ with respect to the basis $\mathbf{e}_1, \mathbf{e}_2, \cdots, \mathbf{e}_m$. Since the c_i satisfy the system (11), this system has a unique solution.

Thus, *the coordinates* c_i *of the orthogonal projection* $\mathbf{f}_0$ *of the vector* $\mathbf{f}$ *on the subspace* $\mathbf{R}_1$ *are determined from the system* (12) *or from the system* (11) *according as the* c_i *are the coordinates of* $\mathbf{f}_0$ *relative to an orthonormal basis of* $\mathbf{R}_1$ *or a non-orthonormal basis of* $\mathbf{R}_1$.

A system of m linear equations in m unknowns can have a unique solution only if its determinant is different from zero. It follows that the determinant of the system (11)

$$\begin{vmatrix} (\mathbf{e}_1, \mathbf{e}_1) & (\mathbf{e}_2, \mathbf{e}_1) & \cdots & (\mathbf{e}_m, \mathbf{e}_1) \\ (\mathbf{e}_1, \mathbf{e}_2) & (\mathbf{e}_2, \mathbf{e}_2) & \cdots & (\mathbf{e}_m, \mathbf{e}_2) \\ \cdots & \cdots & \cdots & \cdots \\ (\mathbf{e}_1, \mathbf{e}_m) & (\mathbf{e}_2, \mathbf{e}_m) & \cdots & (\mathbf{e}_m, \mathbf{e}_m) \end{vmatrix}$$

must be different from zero. This determinant is known as the Gramm determinant of the vectors $\mathbf{e}_1, \mathbf{e}_2, \cdots, \mathbf{e}_m$.

EXAMPLES. *1. The method of least squares.* Let y be a linear function of $x_1, x_2, \cdots, x_m$; i.e., let

$$y = c_1x_1 + \cdots + c_mx_m,$$

where the c_i are fixed unknown coefficients. Frequently the c_i are determined experimentally. To this end one carries out a number of measurements of $x_1, x_2, \cdots, x_m$ and y. Let $x_{1k}, x_{2k}, \cdots, x_{mk}, y_k$ denote the results of the kth measurement. One could try to determine the coefficients $c_1, c_2, \cdots, c_m$ from the system of equations

$$\begin{aligned} x_{11}c_1 + x_{21}c_2 + \cdots + x_{m1}c_m &= y_1, \\ x_{12}c_1 + x_{22}c_2 + \cdots + x_{m2}c_m &= y_2, \\ \cdots\cdots\cdots\cdots\cdots\cdots\cdots\cdots\cdots \\ x_{1n}c_1 + x_{2n}c_2 + \cdots + x_{mn}c_m &= y_n. \end{aligned} \tag{13}$$

However, usually the number n of measurements exceeds the number m of unknowns and the results of the measurements are never free from error. Thus, the system (13) is usually incompatible and can be solved only approximately. There arises the problem of determining $c_1, c_2, \cdots, c_m$ so that the left sides of the equations in (13) are as "close" as possible to the corresponding right sides. As a measure of "closeness" we take the so-called *mean deviation* of the left sides of the equations from the corresponding free terms, i.e., the quantity

$$\sum_{k=1}^{n} (x_{1k}c_1 + x_{2k}c_2 + \cdots + x_{mk}c_k - y_k)^2. \tag{14}$$

The problem of minimizing the mean deviation can be solved directly. However, its solution can be immediately obtained from the results just presented.

Indeed, let us consider the n-dimensional Euclidean space of n-tuples and the following vectors: $\mathbf{e}_1 = (x_{11}, x_{12}, \cdots, x_{1n})$, $\mathbf{e}_2 = (x_{21}, x_{22}, \cdots, x_{2n})$, $\cdots$, $\mathbf{e}_m = (x_{m1}, x_{m2}, \cdots, x_{mn})$, and $f = (y_1, y_2, \cdots, y_n)$ in that space. The right sides of (13) are the components of the vector $\mathbf{f}$ and the left sides, of the vector

$$c_1\mathbf{e}_1 + c_2\mathbf{e}_2 + \cdots + c_m\mathbf{e}_m.$$

Consequently, (14) represents the square of the distance from $\mathbf{f}$ to $c_1\mathbf{e}_1 + c_2\mathbf{e}_2 + \cdots + c_m\mathbf{e}_m$ and the problem of minimizing the mean deviation is equivalent to the problem of choosing m numbers $c_1, c_2, \cdots, c_m$ so as to minimize the distance from $\mathbf{f}$ to $\mathbf{f}_0 = c_1\mathbf{e}_1 + c_2\mathbf{e}_2 + \cdots + c_m\mathbf{e}_m$. If $\mathbf{R}_1$ is the subspace spanned by the vectors $\mathbf{e}_1, \mathbf{e}_2, \cdots, \mathbf{e}_m$ (supposed linearly independent), then our problem is the problem of finding the projection of $\mathbf{f}$ on $\mathbf{R}_1$. As we have seen (cf. formula (11)), the numbers $c_1, c_2, \cdots, c_m$ which solve this problem are found from the system of equations

$$\begin{aligned}(\mathbf{e}_1, \mathbf{e}_1)c_1 + (\mathbf{e}_2, \mathbf{e}_1)c_2 + \cdots + (\mathbf{e}_m, \mathbf{e}_1)c_m &= (\mathbf{f}, \mathbf{e}_1),\\ (\mathbf{e}_1, \mathbf{e}_2)c_1 + (\mathbf{e}_2, \mathbf{e}_2)c_2 + \cdots + (\mathbf{e}_m, \mathbf{e}_2)c_m &= (\mathbf{f}, \mathbf{e}_2),\\ \cdots\cdots\cdots\cdots\cdots\cdots\cdots\cdots\cdots\cdots\cdots\cdots &\\ (\mathbf{e}_1, \mathbf{e}_m)c_1 + (\mathbf{e}_2, \mathbf{e}_m)c_2 + \cdots + (\mathbf{e}_m, \mathbf{e}_m)c_m &= (\mathbf{f}, \mathbf{e}_m),\end{aligned} \tag{15}$$

where

$$(\mathbf{f}, \mathbf{e}_k) = \sum_{j=1}^{n} x_{kj}y_j; \qquad (\mathbf{e}_i, \mathbf{e}_k) = \sum_{j=1}^{n} x_{ij}x_{kj}.$$

The system of equations (15) is referred to as the system of *normal equations*. The method of approximate solution of the system (13) which we have just described is known as the *method of least squares*.

EXERCISE. Use the method of least squares to solve the system of equations

$$\begin{aligned}2c &= 3\\ 3c &= 4\\ 4c &= 5.\end{aligned}$$

Solution: $\mathbf{e}_1 = (2, 3, 4)$, $\mathbf{f} = (3, 4, 5)$. In this case the normal system consists of the single equation

$$(\mathbf{e}_1, \mathbf{e}_1)c = (\mathbf{e}_1, \mathbf{f}),$$

i.e.,

$$29c = 38; \qquad c = 38/29.$$

When the system (13) consists of n equations in one unknown

$$\begin{aligned}x_1 c &= y_1,\\ x_2 c &= y_2,\\ \cdots&\cdots\\ x_n c &= y_n,\end{aligned} \tag{13'}$$

the (least squares) solution is

$$c = \frac{(\mathbf{x}, \mathbf{y})}{(\mathbf{x}, \mathbf{x})} = \frac{\sum_{k=1}^{n} x_k y_k}{\sum_{k=1}^{n} x_k^2}.$$

In this case the geometric significance of c is that of the slope of a line through the origin which is "as close as possible" to the points (x_1, y_1), (x_2, y_2), $\cdots$, (x_n, y_n).

2. *Approximation of functions by means of trigonometric polynomials.* Let $f(t)$ be a continuous function on the interval $[0, 2\pi]$. It is frequently necessary to find a trigonometric polynomial $P(t)$ of given degree which differs from $f(t)$ by as little as possible. We shall measure the proximity of $f(t)$ and $P(t)$ by means of the integral

$$\int_0^{2\pi} [f(t) - P(t)]^2 \, dt. \tag{16}$$

Thus, *we are to find among all trigonometric polynomials of degree n,*

$$P(t) = (a_0/2) + a_1 \cos t + b_1 \sin t + \cdots + a_n \cos nt + b_n \sin nt, \tag{17}$$

that polynomial for which the mean deviation from $f(t)$ is a minimum.

Let us consider the space $\mathbf{R}$ of continuous functions on the interval $[0, 2\pi]$ in which the inner product is defined, as usual, by means of the integral

$$(f, g) = \int_0^{2\pi} f(t)g(t) \, dt.$$

Then the length of a vector $f(t)$ in $\mathbf{R}$ is given by

$$|f| = \sqrt{\int_0^{2\pi} [f(t)]^2 \, dt}.$$

Consequently, the mean deviation (16) is simply the square of the distance from $f(t)$ to $P(t)$. The trigonometric polynomials (17) form a subspace $\mathbf{R}_1$ of $\mathbf{R}$ of dimension $2n + 1$. Our problem is to find that vector of $\mathbf{R}_1$ which is closest to $f(t)$, and this problem is solved by dropping a perpendicular from $f(t)$ to $\mathbf{R}_1$.

Since the functions

$$\mathbf{e}_0 = \frac{1}{\sqrt{2\pi}}; \qquad \mathbf{e}_1 = \frac{\cos t}{\sqrt{\pi}}; \qquad \mathbf{e}_2 = \frac{\sin t}{\sqrt{\pi}}; \qquad \cdots;$$

$$\mathbf{e}_{2n-1} = \frac{\cos nt}{\sqrt{\pi}}; \qquad \mathbf{e}_{2n} = \frac{\sin nt}{\sqrt{\pi}}$$

form an orthonormal basis in $\mathbf{R}_1$ (cf. para. 1, Example 2), the required element $P(t)$ of $\mathbf{R}_1$ is

$$P(t) = \sum_{k=0}^{2n} c_k \mathbf{e}_k, \tag{18}$$

where

$$c_k = (f, \mathbf{e}_k),$$

or

$$c_0 = \frac{1}{\sqrt{2\pi}} \int_0^{2\pi} f(t)dt; \qquad c_{2k-1} = \frac{1}{\sqrt{\pi}} \int_0^{2\pi} f(t) \cos kt\, dt;$$

$$c_{2k} = \frac{1}{\sqrt{\pi}} \int_0^{2\pi} f(t) \sin kt\, dt.$$

Thus, *for the mean deviation of the trigonometric polynomial*

$$P(t) = \frac{a_0}{2} + \sum_{k=1}^{n} a_k \cos kt + b_k \sin kt$$

from $f(t)$ *to be a minimum the coefficients* a_k *and* b_k *must have the values*

$$a_0 = \frac{1}{\pi} \int_0^{2\pi} f(t)\, dt; \qquad a_k = \frac{1}{\pi} \int_0^{2\pi} f(t) \cos kt\, dt;$$

$$b_k = \frac{1}{\pi} \int_0^{2\pi} f(t) \sin kt\, dt.$$

The numbers a_k and b_k defined above are called the *Fourier coefficients* of the function $f(t)$.

3. *Isomorphism of Euclidean spaces.* We have investigated a number of examples of n-dimensional Euclidean spaces. In each of them the word "vector" had a different meaning. Thus in § 2, Example *2*, "vector" stood for an n-tuple of real numbers, in § 2, Example *5*, it stood for a polynomial, etc.

The question arises which of these spaces are fundamentally different and which of them differ only in externals. To be more specific:

DEFINITION 2. *Two Euclidean spaces* $\mathbf{R}$ *and* $\mathbf{R}'$, *are said to be isomorphic if it is possible to establish a one-to-one correspondence* $\mathbf{x} \leftrightarrow \mathbf{x}'$ ($\mathbf{x} \in \mathbf{R}$, $\mathbf{x}' \in \mathbf{R}'$) *such that*

1. *If* $\mathbf{x} \leftrightarrow \mathbf{x}'$ *and* $\mathbf{y} \leftrightarrow \mathbf{y}'$, *then* $\mathbf{x} + \mathbf{y} \leftrightarrow \mathbf{x}' + \mathbf{y}'$, *i.e., if our correspondence associates with* $\mathbf{x} \in \mathbf{R}$ *the vector* $\mathbf{x}' \in \mathbf{R}'$ *and with* $\mathbf{y} \in \mathbf{R}$ *the vector* $\mathbf{y}' \in \mathbf{R}'$, *then it associates with the sum* $\mathbf{x} + \mathbf{y}$ *the sum* $\mathbf{x}' + \mathbf{y}'$.
2. *If* $\mathbf{x} \leftrightarrow \mathbf{x}'$, *then* $\lambda\mathbf{x} \leftrightarrow \lambda\mathbf{x}'$.
3. *If* $\mathbf{x} \leftrightarrow \mathbf{x}'$ *and* $\mathbf{y} \leftrightarrow \mathbf{y}'$, *then* $(\mathbf{x}, \mathbf{y}) = (\mathbf{x}', \mathbf{y}')$; *i.e., the inner products of corresponding pairs of vectors are to have the same value.*

We observe that if in some n-dimensional Euclidean space $\mathbf{R}$ a theorem stated in terms of addition, scalar multiplication and inner multiplication of vectors has been proved, then the same theorem is valid in every Euclidean space $\mathbf{R}'$, isomorphic to the space $\mathbf{R}$. Indeed, if we replaced vectors from $\mathbf{R}$ appearing in the statement and in the proof of the theorem by corresponding vectors from $\mathbf{R}'$, then, in view of the properties 1, 2, 3 of the definition of isomorphism, all arguments would remain unaffected.

The following theorem settles the problem of isomorphism of different Euclidean vector spaces.

THEOREM 2. *All Euclidean spaces of dimension n are isomorphic.*

We shall show that all n-dimensional Euclidean spaces are isomorphic to a selected "standard" Euclidean space of dimension n. This will prove our theorem.

As our standard n-dimensional space $\mathbf{R}'$ we shall take the space of Example *2*, § 2, in which a vector is an n-tuple of real numbers and in which the inner product of two vectors $\mathbf{x}' = (\xi_1, \xi_2, \cdots, \xi_n)$ and $\mathbf{y}' = (\eta_1, \eta_2, \cdots, \eta_n)$ is defined to be

$$(\mathbf{x}', \mathbf{y}') = \xi_1\eta_1 + \xi_2\eta_2 + \cdots + \xi_n\eta_n.$$

Now let $\mathbf{R}$ be any n-dimensional Euclidean space. Let $\mathbf{e}_1$, $\mathbf{e}_2, \cdots, \mathbf{e}_n$ be an orthonormal basis in $\mathbf{R}$ (we showed earlier that every Euclidean space contains such a basis). We associate with the vector

$$\mathbf{x} = \xi_1\mathbf{e}_1 + \xi_2\mathbf{e}_2 + \cdots + \xi_n\mathbf{e}_n$$

in $\mathbf{R}$ the vector

$$\mathbf{x}' = (\xi_1, \xi_2, \cdots, \xi_n)$$

in $\mathbf{R}'$.

We now show that this correspondence is an isomorphism.

The one-to-one nature of this correspondence is obvious. Conditions 1 and 2 are also immediately seen to hold. It remains to prove that our correspondence satisfies condition 3 of the definition of isomorphism, i.e., that the inner products of corresponding pairs of vectors have the same value. Clearly,

$$(\mathbf{x}, \mathbf{y}) = \xi_1\eta_1 + \xi_2\eta_2 + \cdots + \xi_n\eta_n,$$

because of the assumed orthonormality of the $\mathbf{e}_i$. On the other hand, the definition of inner multiplication in $\mathbf{R}'$ states that

$$(\mathbf{x}', \mathbf{y}') = \xi_1\eta_1 + \xi_2\eta_2 + \cdots + \xi_n\eta_n.$$

Thus

$$(\mathbf{x}', \mathbf{y}') = (\mathbf{x}, \mathbf{y});$$

i.e., the inner products of corresponding pairs of vectors have indeed the same value.

This completes the proof of our theorem.

EXERCISE. Prove this theorem by a method analogous to that used in para. 4, § 1.

The following is an interesting consequence of the isomorphism theorem. Any "geometric" assertion (i.e., an assertion stated in terms of addition, inner multiplication and multiplication of vectors by scalars) pertaining to two or three vectors is true if it is true in elementary geometry of three space. Indeed, the vectors in question span a subspace of dimension at most three. This subspace is isomorphic to ordinary three space (or a subspace of it), and it therefore suffices to verify the assertion in the latter space. In particular the Schwarz inequality — a geometric theorem about a pair of vectors — is true in any vector space because it is true in elementary geometry. We thus have a new proof of the Schwarz inequality. Again, inequality (7) of § 2

$$|\mathbf{x} + \mathbf{y}| \leqq |\mathbf{x}| + |\mathbf{y}|,$$

is stated and proved in every textbook of elementary geometry as the proposition that the length of the diagonal of a parallelogram does not exceed the sum of the lengths of its two non-parallel sides, and is therefore valid in every Euclidean space. To illustrate, the inequality,

$$\sqrt{\int_a^b (f(t) + g(t))^2\, dt} \leqq \sqrt{\int_a^b [f(t)]^2\, dt} + \sqrt{\int_a^b [g(t)]^2\, dt},$$

which expresses inequality (7), § 2, in the space of continuous functions on $[a, b]$, is a direct consequence, via the isomorphism theorem, of the proposition of elementary geometry just mentioned.

§ 4. *Bilinear and quadratic forms*

In this section we shall investigate the simplest real valued functions defined on vector spaces.

1. *Linear functions.* Linear functions are the simplest functions defined on vector spaces.

DEFINITION 1. *A linear function (linear form) f is said to be defined on a vector space if with every vector* $\mathbf{x}$ *there is associated a number $f(\mathbf{x})$ so that the following conditions hold:*

1. $f(\mathbf{x} + \mathbf{y}) = f(\mathbf{x}) + f(\mathbf{y})$,
2. $f(\lambda\mathbf{x}) = \lambda f(\mathbf{x})$.

Let $\mathbf{e}_1, \mathbf{e}_2, \cdots, \mathbf{e}_n$ be a basis in an n-dimensional vector space. Since every vector $\mathbf{x}$ can be represented in the form

$$\mathbf{x} = \xi_1\mathbf{e}_1 + \xi_2\mathbf{e}_2 + \cdots + \xi_n\mathbf{e}_n,$$

the properties of a linear function imply that

$$f(\mathbf{x}) = f(\xi_1\mathbf{e}_1 + \xi_2\mathbf{e}_2 + \cdots + \xi_n\mathbf{e}_n) = \xi_1 f(\mathbf{e}_1) + \xi_2 f(\mathbf{e}_2) + \cdots + \xi_n f(\mathbf{e}_n).$$

Thus, *if* $\mathbf{e}_1, \mathbf{e}_2, \cdots, \mathbf{e}_n$ *is a basis of an n-dimensional vector space* $\mathbf{R}$, $\mathbf{x}$ *a vector whose coordinates in the given basis are $\xi_1, \xi_2, \cdots, \xi_n$, and f a linear function defined on* $\mathbf{R}$, *then*

$$f(\mathbf{x}) = a_1\xi_1 + a_2\xi_2 + \cdots + a_n\xi_n, \tag{1}$$

where $f(\mathbf{e}_i) = a_i (i = 1, 2, \cdots, n)$.

The definition of a linear function given above coincides with the definition of a linear function familiar from algebra. What must be remembered, however, is the dependence of the a_i on the choice of a basis. The exact nature of this dependence is easily explained.

Thus let $\mathbf{e}_1, \mathbf{e}_2, \cdots, \mathbf{e}_n$ and $\mathbf{e}'_1, \mathbf{e}'_2, \cdots, \mathbf{e}'_n$ be two bases in $\mathbf{R}$. Let the $\mathbf{e}'_i$ be expressed in terms of the basis vectors $\mathbf{e}_1, \mathbf{e}_2, \cdots, \mathbf{e}_n$ by means of the equations

$$\begin{aligned}
\mathbf{e}'_1 &= \alpha_{11}\mathbf{e}_1 + \alpha_{21}\mathbf{e}_2 + \cdots + \alpha_{n1}\mathbf{e}_n,\\
\mathbf{e}'_2 &= \alpha_{12}\mathbf{e}_1 + \alpha_{22}\mathbf{e}_2 + \cdots + \alpha_{n2}\mathbf{e}_n,\\
&\cdots\cdots\cdots\cdots\cdots\cdots\cdots\cdots\\
\mathbf{e}'_n &= \alpha_{1n}\mathbf{e}_1 + \alpha_{2n}\mathbf{e}_2 + \cdots + \alpha_{nn}\mathbf{e}_n.
\end{aligned}$$

Further, let

$$f(\mathbf{x}) = a_1\xi_1 + a_2\xi_2 + \cdots + a_n\xi_n$$

relative to the basis $\mathbf{e}_1, \mathbf{e}_2, \cdots, \mathbf{e}_n$, and

$$f(\mathbf{x}) = a'_1\xi'_1 + a'_2\xi'_2 + \cdots + a'_n\xi'_n$$

relative to the basis $\mathbf{e}'_1, \mathbf{e}'_2, \cdots, \mathbf{e}'_n$.

Since $a_i = f(\mathbf{e}_i)$ and $a'_k = f(\mathbf{e}'_k)$, it follows that

$$a'_k = f(\alpha_{1k}\mathbf{e}_1 + \alpha_{2k}\mathbf{e}_2 + \cdots + \alpha_{nk}\mathbf{e}_n) = \alpha_{1k}f(\mathbf{e}_1) + \alpha_{2k}f(\mathbf{e}_2) + \cdots + \alpha_{nk}f(\mathbf{e}_n) = \alpha_{1k}a_1 + \alpha_{2k}a_2 + \cdots + \alpha_{nk}a_n.$$

This shows that *the coefficients of a linear form transform under a change of basis like the basis vectors* (or, as it is sometimes said, *cogrediently*).

2. *Bilinear forms.* In what follows an important role is played by bilinear and quadratic forms (functions).

DEFINITION 2. $A(\mathbf{x}; \mathbf{y})$ *is said to be a bilinear function (bilinear form) of the vectors* $\mathbf{x}$ *and* $\mathbf{y}$ *if*

1. *for any fixed* $\mathbf{y}$, $A(\mathbf{x}; \mathbf{y})$ *is a linear function of* $\mathbf{x}$,
2. *for any fixed* $\mathbf{x}$, $A(\mathbf{x}; \mathbf{y})$ *is a linear function of* $\mathbf{y}$.

In other words, noting the definition of a linear function, conditions 1 and 2 above state that

1. $A(\mathbf{x}_1 + \mathbf{x}_2; \mathbf{y}) = A(\mathbf{x}_1; \mathbf{y}) + A(\mathbf{x}_2; \mathbf{y}), A(\lambda\mathbf{x}; \mathbf{y}) = \lambda A(\mathbf{x}; \mathbf{y})$,
2. $A(\mathbf{x}; \mathbf{y}_1 + \mathbf{y}_2) = A(\mathbf{x}; \mathbf{y}_1) + A(\mathbf{x}; \mathbf{y}_2), A(\mathbf{x}; \mu\mathbf{y}) = \mu A(\mathbf{x}; \mathbf{y})$.

EXAMPLES. *1.* Consider the n-dimensional space of n-tuples of real numbers. Let $\mathbf{x} = (\xi_1, \xi_2, \cdots, \xi_n)$, $\mathbf{y} = (\eta_1, \eta_2, \cdots, \eta_n)$, and define

$$\begin{aligned} A(\mathbf{x}; \mathbf{y}) = {} & a_{11}\xi_1\eta_1 + a_{12}\xi_1\eta_2 + \cdots + a_{1n}\xi_1\eta_n \\ & + a_{21}\xi_2\eta_1 + a_{22}\xi_2\eta_2 + \cdots + a_{2n}\xi_2\eta_n \\ & + \cdots\cdots\cdots\cdots\cdots\cdots \\ & + a_{n1}\xi_n\eta_1 + a_{n2}\xi_n\eta_2 + \cdots + a_{nn}\xi_n\eta_n. \end{aligned} \tag{2}$$

$A(\mathbf{x}; \mathbf{y})$ is a bilinear function. Indeed, if we keep $\mathbf{y}$ fixed, i.e., if we regard $\eta_1, \eta_2, \cdots, \eta_n$ as constants, $\sum_{i,k=1}^{n} a_{ik}\xi_i\eta_k$ depends linearly on the ξ_i; $A(\mathbf{x}; \mathbf{y})$ is a linear function of $\mathbf{x} = (\xi_1, \xi_2, \cdots, \xi_n)$. Again, if $\xi_1, \xi_2, \cdots, \xi_n$ are kept constant, $A(\mathbf{x}; \mathbf{y})$ is a linear function of $\mathbf{y}$.

2. Let $K(s, t)$ be a (fixed) continuous function of two variables s, t. Let $\mathbf{R}$ be the space of continuous functions $f(t)$. If we put

$$A(f; g) = \int_a^b \int_a^b K(s, t) f(s) g(t)\, ds\, dt,$$

then $A(f; g)$ is a bilinear function of the vectors f and g. Indeed, the first part of condition 1 of the definition of a bilinear form means, in this case, that the integral of a sum is the sum of the integrals and the second part of condition 1 that the constant λ may be removed from under the integral sign. Conditions 2 have analogous meaning.

If $K(s, t) \equiv 1$, then

$$A(f; g) = \int_a^b \int_a^b f(s) g(t)\, ds\, dt = \int_a^b f(s)\, ds \int_a^b g(t)\, dt,$$

i.e., $A(f; g)$ is the product of the linear functions $\int_a^b f(s)\, ds$ and $\int_a^b g(t)\, dt$.

EXERCISE. Show that if $f(x)$ and $g(y)$ are linear functions, then their product $f(x) \cdot g(y)$ is a bilinear function.

DEFINITION 3. *A bilinear function (bilinear form) is called symmetric if*

$$A(\mathbf{x}; \mathbf{y}) = A(\mathbf{y}; \mathbf{x})$$

for arbitrary vectors $\mathbf{x}$ *and* $\mathbf{y}$.

In Example *1* above the bilinear form $A(\mathbf{x}; \mathbf{y})$ defined by (2) is symmetric if and only if $a_{ik} = a_{ki}$ for all i and k.

The inner product $(\mathbf{x}, \mathbf{y})$ *in a Euclidean space is an example of a symmetric bilinear form.*

Indeed, Axioms 1, 2, 3 in the definition of an inner product (§ 2) say that the inner product is a symmetric, bilinear form.

3. *The matrix of a bilinear form.* We defined a bilinear form axiomatically. Now let $\mathbf{e}_1, \mathbf{e}_2, \cdots, \mathbf{e}_n$ be a basis in n-dimensional space. We shall express the bilinear form $A(\mathbf{x}; \mathbf{y})$ using the coordinates $\xi_1, \xi_2, \cdots, \xi_n$ of $\mathbf{x}$ and the coordinates $\eta_1, \eta_2, \cdots, \eta_n$ of $\mathbf{y}$ relative to the basis $\mathbf{e}_1, \mathbf{e}_2, \cdots, \mathbf{e}_n$. Thus,

$$A(\mathbf{x}; \mathbf{y}) = A(\xi_1 \mathbf{e}_1 + \xi_2 \mathbf{e}_2 + \cdots + \xi_n \mathbf{e}_n; \eta_1 \mathbf{e}_1 + \eta_2 \mathbf{e}_2 + \cdots + \eta_n \mathbf{e}_n).$$

In view of the properties 1 and 2 of bilinear forms

$$A(\mathbf{x}; \mathbf{y}) = \sum_{i,k=1}^{n} A(\mathbf{e}_i; \mathbf{e}_k)\xi_i\eta_k,$$

or, if we denote the constants $A(\mathbf{e}_i; \mathbf{e}_k)$ by a_{ik},

$$A(\mathbf{x}; \mathbf{y}) = \sum_{i,k=1}^{n} a_{ik}\xi_i\eta_k.$$

To sum up: *Every bilinear form in n-dimensional space can be written as*

$$A(\mathbf{x}; \mathbf{y}) = \sum_{i,k=1}^{n} a_{ik}\xi_i\eta_k, \tag{3}$$

where $\mathbf{x} = \xi_1\mathbf{e}_1 + \cdots + \xi_n\mathbf{e}_n$, $\mathbf{y} = \eta_1\mathbf{e}_1 + \cdots + \eta_n\mathbf{e}_n$, *and*

$$a_{ik} = A(\mathbf{e}_i; \mathbf{e}_k). \tag{4}$$

The matrix $\mathscr{A} = ||a_{ik}||$ is called *the matrix of the bilinear form* $A(\mathbf{x}; \mathbf{y})$ *relative to the basis* $\mathbf{e}_1, \mathbf{e}_2, \cdots, \mathbf{e}_n$.

Thus given a basis $\mathbf{e}_1, \mathbf{e}_2, \cdots, \mathbf{e}_n$ the form $A(\mathbf{x}; \mathbf{y})$ is determined by its matrix $\mathscr{A} = ||a_{ik}||$.

EXAMPLE. Let $\mathbf{R}$ be the three-dimensional vector space of triples (ξ_1, ξ_2, ξ_3) of real numbers. We define a bilinear form in $\mathbf{R}$ by means of the equation

$$A(\mathbf{x}; \mathbf{y}) = \xi_1\eta_1 + 2\xi_2\eta_2 + 3\xi_3\eta_3.$$

Let us choose as a basis of $\mathbf{R}$ the vectors

$$\mathbf{e}_1 = (1, 1, 1); \qquad \mathbf{e}_2 = (1, 1, -1); \qquad \mathbf{e}_3 = (1, -1, -1),$$

and compute the matrix $\mathscr{A}$ of the bilinear form $A(\mathbf{x}; \mathbf{y})$. Making use of (4) we find that:

$$\begin{aligned}
a_{11} &= 1 \cdot 1 + 2 \cdot 1 \cdot 1 + 3 \cdot 1 \cdot 1 = 6,\\
a_{12} &= a_{21} = 1 \cdot 1 + 2 \cdot 1 \cdot 1 + 3 \cdot 1 \cdot (-1) = 0,\\
a_{22} &= 1 \cdot 1 + 2 \cdot 1 \cdot 1 + 3 \cdot (-1) \cdot (-1) = 6,\\
a_{13} &= a_{31} = 1 \cdot 1 + 2 \cdot 1 \cdot (-1) + 3 \cdot 1 \cdot (-1) = -4,\\
a_{23} &= a_{32} = 1 \cdot 1 + 2 \cdot 1 \cdot (-1) + 3 \cdot (-1) \cdot (-1) = 2,\\
a_{33} &= 1 \cdot 1 + 2 \cdot (-1) \cdot (-1) + 3 \cdot (-1) \cdot (-1) = 6,
\end{aligned}$$

i.e.,

$$\mathscr{A} = \begin{bmatrix} 6 & 0 & -4 \\ 0 & 6 & 2 \\ -4 & 2 & 6 \end{bmatrix}.$$

It follows that if the coordinates of $\mathbf{x}$ and $\mathbf{y}$ relative to the basis $\mathbf{e}_1, \mathbf{e}_2, \mathbf{e}_3$ are denoted by ξ'_1, ξ'_2, ξ'_3, and $\eta'_1, \eta'_2, \eta'_3$, respectively, then

$$A(\mathbf{x}; \mathbf{y}) = 6\xi'_1\eta'_1 - 4\xi'_1\eta'_3 + 6\xi'_2\eta'_2 + 2\xi'_2\eta'_3 - 4\xi'_3\eta'_1 + 2\xi'_3\eta'_2 + 6\xi'_3\eta'_3.$$

4. *Transformation of the matrix of a bilinear form under a change of basis.* Let $\mathbf{e}_1, \mathbf{e}_2, \cdots, \mathbf{e}_n$ and $\mathbf{f}_1, \mathbf{f}_2, \ldots, \mathbf{f}_n$ be two bases of an n-dimensional vector space. Let the connection between these bases be described by the relations

$$
\begin{aligned}
\mathbf{f}_1 &= c_{11}\mathbf{e}_1 + c_{21}\mathbf{e}_2 + \cdots + c_{n1}\mathbf{e}_n, \\
\mathbf{f}_2 &= c_{12}\mathbf{e}_1 + c_{22}\mathbf{e}_2 + \cdots + c_{n2}\mathbf{e}_n, \\
&\cdots\cdots\cdots\cdots\cdots\cdots\cdots\cdots \\
\mathbf{f}_n &= c_{1n}\mathbf{e}_1 + c_{2n}\mathbf{e}_2 + \cdots + c_{nn}\mathbf{e}_n,
\end{aligned} \tag{5}
$$

which state that the coordinates of the vector $\mathbf{f}_k$ relative to the basis $\mathbf{e}_1, \mathbf{e}_2, \cdots, \mathbf{e}_n$ are $c_{1k}, c_{2k}, \cdots, c_{nk}$. The matrix

$$
\mathscr{C} = \begin{bmatrix} c_{11} c_{12} \cdots c_{1n} \\ c_{21} c_{22} \cdots c_{2n} \\ \cdots\cdots\cdots \\ c_{n1} c_{n2} \cdots c_{nn} \end{bmatrix}
$$

is referred to as the matrix of transition from the basis $\mathbf{e}_1, \mathbf{e}_2, \cdots, \mathbf{e}_n$ to the basis $\mathbf{f}_1, \mathbf{f}_2, \cdots, \mathbf{f}_n$.

Let $\mathscr{A} = ||a_{ik}||$ be the matrix of a bilinear form $A(\mathbf{x}; \mathbf{y})$ relative to the basis $\mathbf{e}_1, \mathbf{e}_2, \cdots, \mathbf{e}_n$ and $\mathscr{B} = ||b_{ik}||$, the matrix of that form relative to the basis $\mathbf{f}_1, \mathbf{f}_2, \cdots, \mathbf{f}_n$. Our problem consists in finding the matrix $||b_{ik}||$ given the matrix $||a_{ik}||$.

By definition [eq. (4)] $b_{pq} = A(\mathbf{f}_p; \mathbf{f}_q)$, i.e., b_{pq} is the value of our bilinear form for $\mathbf{x} = \mathbf{f}_p$, $\mathbf{y} = \mathbf{f}_q$. To find this value we make use of (3) where in place of the ξ_i and η_j we put the coordinates of $\mathbf{f}_p$ and $\mathbf{f}_q$ relative to the basis $\mathbf{e}_1, \mathbf{e}_2, \cdots, \mathbf{e}_n$, i.e., the numbers $c_{1p}, c_{2p}, \cdots, c_{np}$ and $c_{1q}, c_{2q}, \cdots, c_{nq}$. It follows that

$$
b_{pq} = A(\mathbf{f}_p; \mathbf{f}_q) = \sum_{i,k=1}^{n} a_{ik} c_{ip} c_{kq}. \tag{6}
$$

We shall now express our result in matrix form. To this end we put $c_{ip} = c'_{pi}$. The c'_{pi} are, of course, the elements of the transpose $\mathscr{C}'$ of $\mathscr{C}$. Now b_{pq} becomes [4]

[4] As is well known, the element c_{ik} of a matrix $\mathscr{C}$ which is the product of two matrices $\mathscr{A} = ||a_{ik}||$ and $\mathscr{B} = ||b_{ik}||$ is defined as

$$
c_{ik} = \sum_{\alpha=1}^{n} a_{i\alpha} b_{\alpha k}.
$$

Using this definition twice one can show that if $\mathscr{D} = \mathscr{A}\mathscr{B}\mathscr{C}$, then

$$
d_{ik} = \sum_{\alpha, \beta=1}^{n} a_{i\alpha} b_{\alpha\beta} c_{\beta k}.
$$

$$(7^*) \qquad b_{pq} = \sum_{i,k=1}^{n} c'_{pi} a_{ik} c_{kq}.$$

Using matrix notation we can state that

$$(7) \qquad \mathscr{B} = \mathscr{C}' \mathscr{A} \mathscr{C}.$$

Thus, if $\mathscr{A}$ *is the matrix of a bilinear form* $A(\mathbf{x}; \mathbf{y})$ *relative to the basis* $\mathbf{e}_1, \mathbf{e}_2, \cdots, \mathbf{e}_n$ *and* $\mathscr{B}$ *its matrix relative to the basis* $\mathbf{f}_1, \mathbf{f}_2, \cdots, \mathbf{f}_n$, *then* $\mathscr{B} = \mathscr{C}' \mathscr{A} \mathscr{C}$, *where* $\mathscr{C}$ *is the matrix of transition from* $\mathbf{e}_1, \mathbf{e}_2, \cdots, \mathbf{e}_n$ *to* $\mathbf{f}_1, \mathbf{f}_2, \cdots, \mathbf{f}_n$ *and* $\mathscr{C}'$ *is the transpose of* $\mathscr{C}$.

5. *Quadratic forms*

DEFINITION 4. *Let* $A(\mathbf{x}; \mathbf{y})$ *be a symmetric bilinear form. The function* $A(\mathbf{x}; \mathbf{x})$ *obtained from* $A(\mathbf{x}; \mathbf{y})$ *by putting* $\mathbf{y} = \mathbf{x}$ *is called a quadratic form.*

$A(\mathbf{x}; \mathbf{y})$ is referred to as the bilinear form *polar* to the quadratic form $A(\mathbf{x}; \mathbf{x})$.

The requirement of Definition 4 that $A(\mathbf{x}; \mathbf{y})$ be a symmetric form is justified by the following result which would be invalid if this requirement were dropped.

THEOREM 1. *The polar form* $A(\mathbf{x}; \mathbf{y})$ *is uniquely determined by its quadratic form.*

Proof: The definition of a bilinear form implies that

$$A(\mathbf{x} + \mathbf{y}; \mathbf{x} + \mathbf{y}) = A(\mathbf{x}; \mathbf{x}) + A(\mathbf{x}; \mathbf{y}) + A(\mathbf{y}; \mathbf{x}) + A(\mathbf{y}; \mathbf{y}).$$

Hence in view of the symmetry of $A(\mathbf{x}; \mathbf{y})$ (i.e., in view of the equality $A(\mathbf{x}; \mathbf{y}) = A(\mathbf{y}; \mathbf{x})$),

$$A(\mathbf{x}; \mathbf{y}) = \tfrac{1}{2}[A(\mathbf{x} + \mathbf{y}; \mathbf{x} + \mathbf{y}) - A(\mathbf{x}; \mathbf{x}) - A(\mathbf{y}; \mathbf{y})].$$

Since the right side of the above equation involves only values of the quadratic form $A(\mathbf{x}; \mathbf{x})$, it follows that $A(\mathbf{x}; \mathbf{y})$ is indeed uniquely determined by $A(\mathbf{x}; \mathbf{x})$.

To show the essential nature of the symmetry requirement in the above result we need only observe that if $A(\mathbf{x}; \mathbf{y})$ is any (not necessarily symmetric) bilinear form, then $A(\mathbf{x}; \mathbf{y})$ as well as the symmetric bilinear form

$$A_1(\mathbf{x}; \mathbf{y}) = \tfrac{1}{2}[A(\mathbf{x}; \mathbf{y}) + A(\mathbf{y}; \mathbf{x})]$$

give rise to the same quadratic form $A(\mathbf{x}; \mathbf{x})$.

We have already shown that every symmetric bilinear form $A(\mathbf{x}; \mathbf{y})$ can be expressed in terms of the coordinates ξ_i of $\mathbf{x}$ and η_k of $\mathbf{y}$ as follows:

$$A(\mathbf{x}; \mathbf{y}) = \sum_{i, k=1}^{n} a_{ik}\xi_i\eta_k,$$

where $a_{ik} = a_{ki}$. It follows that *relative to a given basis every quadratic form* $A(\mathbf{x}; \mathbf{x})$ *can be expressed as follows:*

$$A(\mathbf{x}; \mathbf{x}) = \sum_{i, k=1}^{n} a_{ik}\xi_i\xi_k, \quad a_{ik} = a_{ki}.$$

We introduce another important

DEFINITION 5. *A quadratic form* $A(\mathbf{x}; \mathbf{x})$ *is called positive definite if for every vector* $\mathbf{x} \neq \mathbf{0}$

$$A(\mathbf{x}; \mathbf{x}) > 0.$$

EXAMPLE. It is clear that $A(\mathbf{x}; \mathbf{x}) = \xi_1^2 + \xi_2^2 + \cdots + \xi_n^2$ is a positive definite quadratic form.

Let $A(\mathbf{x}; \mathbf{x})$ be a positive definite quadratic form and $A(\mathbf{x}; \mathbf{y})$ its polar form. The definitions formulated above imply that

1. $A(\mathbf{x}; \mathbf{y}) = A(\mathbf{y}; \mathbf{x})$.
2. $A(\mathbf{x}_1 + \mathbf{x}_2; \mathbf{y}) = A(\mathbf{x}_1; \mathbf{y}) + A(\mathbf{x}_2; \mathbf{y})$.
3. $A(\lambda\mathbf{x}; \mathbf{y}) = \lambda A(\mathbf{x}; \mathbf{y})$.
4. $A(\mathbf{x}; \mathbf{x}) \geqq 0$ and $A(\mathbf{x}; \mathbf{x}) > 0$ for $\mathbf{x} \neq \mathbf{0}$.

These conditions are seen to coincide with the axioms for an inner product stated in § 2. Hence,

an inner product is a bilinear form corresponding to a positive definite quadratic form. Conversely, such a bilinear form always defines an inner product.

This enables us to give the following alternate definition of Euclidean space:

A vector space is called Euclidean if there is defined in it a positive definite quadratic form $A(\mathbf{x}; \mathbf{x})$. *In such a space the value of the inner product* $(\mathbf{x}, \mathbf{y})$ *of two vectors is taken as the value* $A(\mathbf{x}; \mathbf{y})$ *of the (uniquely determined) bilinear form* $A(\mathbf{x}; \mathbf{y})$ *associated with* $A(\mathbf{x}; \mathbf{x})$.

§ 5. *Reduction of a quadratic form to a sum of squares*

We know by now that the expression for a quadratic form $A(\mathbf{x}; \mathbf{x})$ in terms of the coordinates of the vector $\mathbf{x}$ depends on the choice of basis. We now show how to select a basis (coordinate system) in which the quadratic form is represented as a sum of squares, i.e.,

$$(1) \qquad A(\mathbf{x}; \mathbf{x}) = \lambda_1\xi_1^2 + \lambda_2\xi_2^2 + \cdots + \lambda_n\xi_n^2.$$

Thus let $\mathbf{f}_1, \mathbf{f}_2, \cdots, \mathbf{f}_n$ be a basis of our space and let

$$(2) \qquad A(\mathbf{x}; \mathbf{x}) = \sum_{i,k=1}^{n} a_{ik}\eta_i\eta_k,$$

where $\eta_1, \eta_2, \cdots, \eta_n$ are the coordinates of the vector $\mathbf{x}$ relative to this basis. We shall now carry out a succession of basis transformations aimed at eliminating the terms in (2) containing products of coordinates with different indices. In view of the one-to-one correspondence between coordinate transformations and basis transformations (cf. para. 6, § 1) we may write the formulas for coordinate transformations in place of formulas for basis transformations.

To reduce the quadratic form $A(\mathbf{x}; \mathbf{x})$ to a sum of squares it is necessary to begin with an expression (2) for $A(\mathbf{x}; \mathbf{x})$ in which at least one of the a_{kk} (a_{kk} is the coefficient of η_k^2) is not zero. If the form $A(\mathbf{x}; \mathbf{x})$ (supposed not identically zero) does not contain any square of the variables $\eta_1, \eta_2, \cdots, \eta_n$, it contains one product say, $2a_{12}\eta_1\eta_2$. Consider the coordinate transformation defined by

$$\begin{aligned} \eta_1 &= \eta'_1 + \eta'_2 \\ \eta_2 &= \eta'_1 - \eta'_2 \qquad\qquad (k = 3, \cdots, n) \\ \eta_k &= \eta'_k \end{aligned}$$

Under this transformation $2a_{12}\eta_1\eta_2$ goes over into $2a_{12}(\eta'^2_1 - \eta'^2_2)$. Since $a_{11} = a_{22} = 0$, the coefficient of η'^2_1 stays different from zero.

We shall assume slightly more than we may on the basis of the above, namely, that in (2) $a_{11} \neq 0$. If this is not the case it can be brought about by a change of basis consisting in a suitable change of the numbering of the basis elements. We now single out all those terms of the form which contain η_1

$$a_{11}\eta_1^2 + 2a_{12}\eta_1\eta_2 + \cdots + 2a_{1n}\eta_1\eta_n,$$

and "complete the square," i.e., write

$$
\begin{aligned}
&a_{11}\eta_1^2 + 2a_{12}\eta_1\eta_2 + \cdots + 2a_{1n}\eta_1\eta_n \\
(3) \qquad &= \frac{1}{a_{11}}(a_{11}\eta_1 + \cdots + a_{1n}\eta_n)^2 - B.
\end{aligned}
$$

It is clear that B contains only squares and products of the terms $a_{12}\eta_2, \cdots, a_{1n}\eta_n$ so that upon substitution of the right side of (3) in (2) the quadratic form under consideration becomes

$$A(\mathbf{x}; \mathbf{x}) = \frac{1}{a_{11}}(a_{11}\eta_1 + \cdots + a_{1n}\eta_n)^2 + \cdots,$$

where the dots stand for a sum of terms in the variables $\eta_2, \cdots \eta_n$.

If we put

$$
\begin{aligned}
\eta_1^* &= a_{11}\eta_1 + a_{12}\eta_2 + \cdots + a_{1n}\eta_n, \\
\eta_2^* &= \qquad\qquad \eta_2, \\
&\cdots\cdots\cdots\cdots\cdots\cdots\cdots \\
\eta_n^* &= \qquad\qquad\qquad\qquad \eta_n,
\end{aligned}
$$

then our quadratic form goes over into

$$A(\mathbf{x}; \mathbf{x}) = \frac{1}{a_{11}}\eta_1^{*2} + \sum_{i,k=2}^{n} a_{ik}^*\eta_i^*\eta_k^*.$$

The expression $\sum_{i,k=2}^{n} a_{ik}^*\eta_i^*\eta_k^*$ is entirely analogous to the right side of (2) except for the fact that it does not contain the first coordinate. If we assume that $a_{22}^* \neq 0$ (which can be achieved, if necessary, by auxiliary transformations discussed above) and carry out another change of coordinates defined by

$$
\begin{aligned}
\eta_1^{**} &= \eta_1^*, \\
\eta_2^{**} &= a_{22}^*\eta_2^* + a_{23}^*\eta_3^* + \cdots + a_{2n}^*\eta_n^*, \\
\eta_3^{**} &= \qquad\qquad \eta_3^*, \\
&\cdots\cdots\cdots\cdots\cdots\cdots\cdots \\
\eta_n^{**} &= \qquad\qquad\qquad\qquad \eta_n^*,
\end{aligned}
$$

our form becomes

$$A(\mathbf{x}; \mathbf{x}) = \frac{1}{a_{11}}\eta_1^{**2} + \frac{1}{a_{22}^*}\eta_2^{**2} + \sum_{i,k=3}^{n} a_{ik}^{**}\eta_i^{**}\eta_k^{**}.$$

After a finite number of steps of the type just described our expression will finally take the form

$$A(\mathbf{x}; \mathbf{x}) = \lambda_1\xi_1^2 + \lambda_2\xi_2^2 + \cdots + \lambda_m\xi_m^2,$$

where $m \leqq n$.

We leave it as an exercise for the reader to write out the basis transformation corresponding to each of the coordinate transformations utilized in the process of reduction of $A(\mathbf{x}; \mathbf{x})$ (cf. para. 6, § 1) and to see that each change leads from basis to basis, i.e., to n linearly independent vectors.

If $m < n$, we put $\lambda_{m+1} = \cdots = \lambda_n = 0$. We may now sum up our conclusions as follows:

THEOREM 1. *Let $A(\mathbf{x}; \mathbf{x})$ be a quadratic form in an n-dimensional space* **R**. *Then there exists a basis $\mathbf{e}_1, \mathbf{e}_2, \cdots, \mathbf{e}_n$ of* **R** *relative to which $A(\mathbf{x}; \mathbf{x})$ has the form*

$$A(\mathbf{x}; \mathbf{x}) = \lambda_1\xi_1^2 + \lambda_2\xi_2^2 + \cdots + \lambda_n\xi_n^2,$$

where $\xi_1, \xi_2, \cdots, \xi_n$ are the coordinates of $\mathbf{x}$ *relative to* $\mathbf{e}_1, \mathbf{e}_2, \cdots, \mathbf{e}_n$.

We shall now give an example illustrating the above method of reducing a quadratic form to a sum of squares. Thus let $A(\mathbf{x}; \mathbf{x})$ be a quadratic form in three-dimensional space which is defined, relative to some basis $\mathbf{f}_1, \mathbf{f}_2, \mathbf{f}_3$, by the equation

$$A(\mathbf{x}; \mathbf{x}) = 2\eta_1\eta_2 + 4\eta_1\eta_3 - \eta_2^2 - 8\eta_3^2.$$

If

$$\begin{aligned} \eta_1 &= \eta'_2, \\ \eta_2 &= \eta'_1, \\ \eta_3 &= \eta'_3, \end{aligned}$$

then

$$A(\mathbf{x}; \mathbf{x}) = -\eta'^2_1 + 2\eta'_1\eta'_2 + 4\eta'_2\eta'_3 - 8\eta'^2_3.$$

Again, if

$$\begin{aligned} \eta_1^* &= -\eta'_1 + \eta'_2 \\ \eta_2^* &= \qquad\quad \eta'_2, \\ \eta_3^* &= \qquad\quad \eta'_3, \end{aligned}$$

then

$$A(\mathbf{x}; \mathbf{x}) = -\eta_1^{*2} + \eta_2^{*2} + 4\eta_2^*\eta_3^* - 8\eta_3^{*2}.$$

Finally, if

$$\begin{aligned} \xi_1 &= \eta_1^*, \\ \xi_2 &= \qquad \eta_2^* + 2\eta_3^*, \\ \xi_3 &= \qquad\qquad \eta_3^*, \end{aligned}$$

then $A(\mathbf{x}; \mathbf{x})$ assumes the canonical form

$$A(\mathbf{x}; \mathbf{x}) = -\xi_1^2 + \xi_2^2 - 12\xi_3^2.$$

If we have the expressions for $\eta_1^*, \eta_2^*, \cdots, \eta_n^*$ in terms of $\eta_1, \eta_2, \cdots, \eta_n$, for $\eta_1^{**}, \eta_2^{**}, \cdots, \eta_n^{**}$ in terms of $\eta_1^*, \eta_2^*, \cdots, \eta_n^*$, etc., we can express $\xi_1, \xi_2, \cdots, \xi_n$ in terms of $\eta_1, \eta_1, \cdots, \eta_n$ in the form

$$\begin{aligned}
\xi_1 &= c_{11}\eta_1 + c_{12}\eta_2 + \cdots + c_{1n}\eta_n \\
\xi_2 &= c_{21}\eta_1 + c_{22}\eta_2 + \cdots + c_{2n}\eta_n \\
&\cdots\cdots\cdots\cdots\cdots\cdots\cdots\cdots \\
\xi_n &= c_{n1}\eta_1 + c_{n2}\eta_2 + \cdots + c_{nn}\eta_n.
\end{aligned}$$

Thus in the example just given

$$\begin{aligned}
\xi_1 &= \eta_1 - \eta_2, \\
\xi_2 &= \eta_1 \qquad\quad + 2\eta_3, \\
\xi_3 &= \qquad\qquad\quad\ \ \eta_3.
\end{aligned}$$

In view of the fact that the matrix of a coordinate transformation is the inverse of the transpose of the matrix of the corresponding basis transformation (cf. para. 6, § 1) we can express the new basis vectors $\mathbf{e}_1, \mathbf{e}_2, \cdots, \mathbf{e}_n$ in terms of the old basis vectors $\mathbf{f}_1, \mathbf{f}_2, \cdots, \mathbf{f}_n$

$$\begin{aligned}
\mathbf{e}_1 &= d_{11}\mathbf{f}_1 + d_{12}\mathbf{f}_2 + \cdots + d_{1n}\mathbf{f}_n \\
\mathbf{e}_2 &= d_{21}\mathbf{f}_1 + d_{22}\mathbf{f}_2 + \cdots + d_{2n}\mathbf{f}_n \\
&\cdots\cdots\cdots\cdots\cdots\cdots\cdots\cdots \\
\mathbf{e}_n &= d_{n1}\mathbf{f}_1 + d_{n2}\mathbf{f}_2 + \cdots + d_{nn}\mathbf{f}_n.
\end{aligned}$$

If the form $A(\mathbf{x}; \mathbf{x})$ is such that at no stage of the reduction process is there need to "create squares" or to change the numbering of the basis elements (cf. the beginning of the description of the reduction process in this section), then the expressions for $\xi_1, \xi_2, \cdots, \xi_n$ in terms of $\eta_1, \eta_2, \cdots, \eta_n$ take the form

$$\begin{aligned}
\xi_1 &= c_{11}\eta_1 + c_{12}\eta_2 + \cdots + c_{1n}\eta_n, \\
\xi_2 &= \qquad\quad\ \ c_{22}\eta_2 + \cdots + c_{2n}\eta_n, \\
&\cdots\cdots\cdots\cdots\cdots\cdots\cdots\cdots \\
\xi_n &= \qquad\qquad\qquad\qquad\quad\ \ c_{nn}\eta_n,
\end{aligned}$$

i.e. the matrix of the coordinate transformation is a so called triangular matrix. It is easy to check that in this case the matrix of the corresponding basis transformation is also a triangular matrix:

$$\begin{aligned} \mathbf{e}_1 &= d_{11}\mathbf{f}_1, \\ \mathbf{e}_2 &= d_{21}\mathbf{f}_1 + d_{22}\mathbf{f}_2, \\ &\cdots\cdots\cdots\cdots\cdots \\ \mathbf{e}_n &= d_{n1}\mathbf{f}_1 + d_{n2}\mathbf{f}_2 + \cdots + d_{nn}\mathbf{f}_n. \end{aligned}$$

§ 6. *Reduction of a quadratic form by means of a triangular transformation*

1. In this section we shall describe another method of constructing a basis in which the quadratic form becomes a sum of squares. In contradistinction to the preceding section we shall express the vectors of the desired basis directly in terms of the vectors of the initial basis. However, this time we shall find it necessary to impose certain restrictions on the form $A(\mathbf{x}; \mathbf{y})$ and the initial basis $\mathbf{f}_1, \mathbf{f}_2, \cdots, \mathbf{f}_n$. Thus let $||a_{ik}||$ be the matrix of the bilinear form $A(\mathbf{x}; \mathbf{y})$ relative to the basis $\mathbf{f}_1, \mathbf{f}_2, \cdots, \mathbf{f}_n$. We assume that the following determinants are different from zero:

$$\Delta_1 = a_{11} \neq 0; \qquad \Delta_2 = \begin{vmatrix} a_{11} & a_{12} \\ a_{21} & a_{22} \end{vmatrix} \neq 0; \qquad \cdots;$$

(1)

$$\Delta_n = \begin{vmatrix} a_{11} & a_{12} & \cdots & a_{1n} \\ a_{21} & a_{22} & \cdots & a_{2n} \\ \cdots & \cdots & \cdots & \cdots \\ a_{n1} & a_{n2} & \cdots & a_{nn} \end{vmatrix} \neq 0.$$

(It is worth noting that this requirement is equivalent to the requirement that in the method of reducing a quadratic form to a sum of squares described in § 5 the coefficients a_{11}, a_{22}^*, etc., be different from zero.

Now let the quadratic form $A(\mathbf{x}; \mathbf{x})$ be defined relative to the basis $\mathbf{f}_1, \mathbf{f}_2, \cdots, \mathbf{f}_n$ by the equation

$$A(\mathbf{x}; \mathbf{x}) = \sum_{i,k=1}^{n} a_{ik}\xi_i\xi_k, \qquad \text{where } a_{ik} = A(\mathbf{f}_i; \mathbf{f}_k).$$

It is our aim to define vectors $\mathbf{e}_1, \mathbf{e}_2, \cdots, \mathbf{e}_n$ so that

(2) $$A(\mathbf{e}_i; \mathbf{e}_k) = 0 \qquad \text{for } i \neq k \ (i, k = 1, 2, \cdots, n).$$

We shall seek these vectors in the form

$$
\begin{aligned}
\mathbf{e}_1 &= \alpha_{11}\mathbf{f}_1, \\
\mathbf{e}_2 &= \alpha_{21}\mathbf{f}_1 + \alpha_{22}\mathbf{f}_2, \\
&\cdots\cdots\cdots\cdots\cdots \\
\mathbf{e}_n &= \alpha_{n1}\mathbf{f}_1 + \alpha_{n2}\mathbf{f}_2 + \cdots + \alpha_{nn}\mathbf{f}_n.
\end{aligned} \tag{3}
$$

We could now determine the coefficients α_{ik} from the conditions (2) by substituting for each vector in (2) the expression for that vector in (3). However, this scheme leads to equations of degree two in the α_{ik} and to obviate the computational difficulties involved we adopt a different approach.

We observe that if

$$A(\mathbf{e}_k;\, \mathbf{f}_i) = 0 \qquad \text{for } i = 1, 2, \cdots, k-1,$$

then

$$A(\mathbf{e}_k;\, \mathbf{e}_i) = 0 \qquad \text{for } i = 1, 2, \cdots, k-1.$$

Indeed, if we replace $\mathbf{e}_i$ by

$$\alpha_{i1}\mathbf{f}_1 + \alpha_{i2}\mathbf{f}_2 + \cdots + \alpha_{ii}\mathbf{f}_i,$$

then

$$
\begin{aligned}
A(\mathbf{e}_k;\ \mathbf{e}_i) &= A(\mathbf{e}_k;\, \alpha_{i1}\mathbf{f}_1 + \alpha_{i2}\mathbf{f}_2 + \cdots + \alpha_{ii}\mathbf{f}_i) \\
&= \alpha_{i1}A(\mathbf{e}_k;\, \mathbf{f}_1) + \alpha_{i2}A(\mathbf{e}_k;\, \mathbf{f}_2) + \cdots + \alpha_{ii}A(\mathbf{e}_k;\, \mathbf{f}_i).
\end{aligned}
$$

Thus if $A(\mathbf{e}_k;\, \mathbf{f}_i) = 0$ for every k and for all $i < k$, then $A(\mathbf{e}_k;\, \mathbf{e}_i) = 0$ for $i < k$ and therefore, in view of the symmetry of the bilinear form, also for $i > k$, i.e., $\mathbf{e}_1, \mathbf{e}_2, \cdots, \mathbf{e}_n$ is the required basis. Our problem then is to find coefficients $\alpha_{k1}, \alpha_{k2}, \cdots, \alpha_{kk}$ such that the vector

$$\mathbf{e}_k = \alpha_{k1}\mathbf{f}_1 + \alpha_{k2}\mathbf{f}_2 + \cdots + \alpha_{kk}\mathbf{f}_k$$

satisfies the relations

$$A(\mathbf{e}_k;\, \mathbf{f}_i) = 0, \qquad (i = 1, 2, \cdots, k-1). \tag{4}$$

We assert that conditions (4) determine the vector $\mathbf{e}_k$ to within a constant multiplier. To fix this multiplier we add the condition

$$A(\mathbf{e}_k;\, \mathbf{f}_k) = 1. \tag{5}$$

We claim that conditions (4) and (5) determine the vector $\mathbf{e}_k$

uniquely. The proof is immediate. Substituting in (4) and (5) the expression for $\mathbf{e}_k$ we are led to the following linear system for the α_{ki}:

$$
(6)\qquad
\begin{aligned}
&\alpha_{k1}A(\mathbf{f}_1;\mathbf{f}_1) + \alpha_{k2}A(\mathbf{f}_1;\mathbf{f}_2) + \cdots + \alpha_{kk}A(\mathbf{f}_1;\mathbf{f}_k) = 0,\\
&\alpha_{k1}A(\mathbf{f}_2;\mathbf{f}_1) + \alpha_{k2}A(\mathbf{f}_2;\mathbf{f}_2) + \cdots + \alpha_{kk}A(\mathbf{f}_2;\mathbf{f}_k) = 0,\\
&\cdots\cdots\cdots\cdots\cdots\cdots\cdots\cdots\cdots\cdots\\
&\alpha_{k1}A(\mathbf{f}_{k-1};\mathbf{f}_1) + \alpha_{k2}A(\mathbf{f}_{k-1};\mathbf{f}_2) + \cdots + \alpha_{kk}A(\mathbf{f}_{k-1};\mathbf{f}_k) = 0,\\
&\alpha_{k1}A(\mathbf{f}_k;\mathbf{f}_1) + \alpha_{k2}A(\mathbf{f}_k;\mathbf{f}_2) + \cdots + \alpha_{kk}A(\mathbf{f}_k;\mathbf{f}_k) = 1.
\end{aligned}
$$

The determinant of this system is equal to

$$
(7)\qquad \Delta_k = \begin{vmatrix}
A(\mathbf{f}_1;\mathbf{f}_1) & A(\mathbf{f}_1,\mathbf{f}_2) & \cdots & A(\mathbf{f}_1;\mathbf{f}_k)\\
A(\mathbf{f}_2;\mathbf{f}_1) & A(\mathbf{f}_2,\mathbf{f}_2) & \cdots & A(\mathbf{f}_2;\mathbf{f}_k)\\
\cdots & \cdots & \cdots & \cdots\\
A(\mathbf{f}_k;\mathbf{f}_1) & A(\mathbf{f}_k,\mathbf{f}_2) & \cdots & A(\mathbf{f}_k;\mathbf{f}_k)
\end{vmatrix}
$$

and is by assumption (1) different from zero so that the system (6) has a unique solution. Thus conditions (4) and (5) determine $\mathbf{e}_k$ uniquely, as asserted.

It remains to find the coefficients b_{ik} of the quadratic form $A(\mathbf{x};\mathbf{x})$ relative to the basis $\mathbf{e}_1, \mathbf{e}_2, \cdots, \mathbf{e}_n$ just constructed. As we already know

$$b_{ik} = A(\mathbf{e}_i;\mathbf{e}_k).$$

The basis of the $\mathbf{e}_i$ is characterized by the fact that $A(\mathbf{e}_i;\mathbf{e}_k) = 0$ for $i \neq k$, i.e., $b_{ik} = 0$ for $i \neq k$. It therefore remains to compute $b_{kk} = A(\mathbf{e}_k;\mathbf{e}_k)$. Now

$$
\begin{aligned}
A(\mathbf{e}_k;\mathbf{e}_k) &= A(\mathbf{e}_k;\alpha_{k1}\mathbf{f}_1 + \alpha_{k2}\mathbf{f}_2 + \cdots + \alpha_{kk}\mathbf{f}_k)\\
&= \alpha_{k1}A(\mathbf{e}_k;\mathbf{f}_1) + \alpha_{k2}A(\mathbf{e}_k;\mathbf{f}_2) + \cdots + \alpha_{kk}A(\mathbf{e}_k;\mathbf{f}_k),
\end{aligned}
$$

which in view of (4) and (5) is the same as

$$A(\mathbf{e}_k;\mathbf{e}_k) = \alpha_{kk}.$$

The number α_{kk} can be found from the system (6). Namely, by Cramer's rule,

$$\alpha_{kk} = \frac{\Delta_{k-1}}{\Delta_k},$$

where Δ_{k-1} is a determinant of order $k-1$ analogous to (7) and $\Delta_0 = 1$.

Thus

$$b_{kk} = A(\mathbf{e}_k; \mathbf{e}_k) = \frac{\Delta_{k-1}}{\Delta_k}.$$

To sum up:

THEOREM 1. *Let $A(\mathbf{x}; \mathbf{x})$ be a quadratic form defined relative to some basis $\mathbf{f}_1, \mathbf{f}_2, \cdots, \mathbf{f}_n$ by the equation*

$$A(\mathbf{x}; \mathbf{x}) = \sum_{i,k=1}^{n} a_{ik}\eta_i\eta_k, \qquad a_{ik} = A(\mathbf{f}_i; \mathbf{f}_k).$$

Further, let the determinants

$$\Delta_1 = a_{11}, \qquad \Delta_2 = \begin{vmatrix} a_{11} & a_{12} \\ a_{21} & a_{22} \end{vmatrix}, \qquad \cdots,$$

$$\Delta_n = \begin{vmatrix} a_{11} & a_{12} & \cdots & a_{1n} \\ a_{21} & a_{22} & \cdots & a_{2n} \\ \cdots & \cdots & \cdots & \cdots \\ a_{n1} & a_{n2} & \cdots & a_{nn} \end{vmatrix}$$

be all different from zero. Then there exists a basis $\mathbf{e}_1, \mathbf{e}_2, \cdots, \mathbf{e}_n$ relative to which $A(\mathbf{x}; \mathbf{x})$ is expressed as a sum of squares,

$$A(\mathbf{x}; \mathbf{x}) = \frac{\Delta_0}{\Delta_1}\xi_1^2 + \frac{\Delta_1}{\Delta_2}\xi_2^2 + \cdots + \frac{\Delta_{n-1}}{\Delta_n}\xi_n^2.$$

Here $\xi_1, \xi_2, \cdots, \xi_n$ are the coordinates of $\mathbf{x}$ in the basis $\mathbf{e}_1, \mathbf{e}_2, \cdots, \mathbf{e}_n$.

This method of reducing a quadratic form to a sum of squares is known as the *method of Jacobi*.

REMARK: The fact that in the proof of the above theorem we were led to a definite basis $\mathbf{e}_1, \mathbf{e}_2, \cdots, \mathbf{e}_n$ in which the quadratic form is expressed as a sum of squares does not mean that this basis is unique. In fact, if one were to start out with another basis $\mathbf{f}_1, \mathbf{f}_2, \cdots, \mathbf{f}_n$ (or if one were simply to permute the vectors $\mathbf{f}_1$, $\mathbf{f}_2, \cdots, \mathbf{f}_n$) one would be led to another basis $\mathbf{e}_1, \mathbf{e}_2, \cdots, \mathbf{e}_n$. Also, it should be pointed out that the vectors $\mathbf{e}_1, \mathbf{e}_2, \cdots, \mathbf{e}_n$ need not have the form (3).

EXAMPLE. Consider the quadratic form

$$2\xi_1^2 + 3\xi_1\xi_2 + 4\xi_1\xi_3 + \xi_2^2 + \xi_3^2$$

in three-dimensional space with basis

$$\mathbf{f}_1 = (1, 0, 0), \quad \mathbf{f}_2 = (0, 1, 0), \quad \mathbf{f}_3 = (0, 0, 1).$$

The corresponding bilinear form is

$$A(\mathbf{x}; \mathbf{y}) = 2\xi_1\eta_1 + \tfrac{3}{2}\xi_1\eta_2 + 2\xi_1\eta_3 + \tfrac{3}{2}\xi_2\eta_1 + \xi_2\eta_2 + 2\xi_3\eta_1 + \xi_3\eta_3.$$

The determinants Δ_1, Δ_2, Δ_3 are 2, $-\frac{1}{4}$, $-\frac{17}{4}$, i.e., none of them vanishes. Thus our theorem may be applied to the quadratic form at hand. Let

$$\begin{aligned} \mathbf{e}_1 &= \alpha_{11}\mathbf{f}_1 &&= (\alpha_{11}, 0, 0), \\ \mathbf{e}_2 &= \alpha_{21}\mathbf{f}_1 + \alpha_{22}\mathbf{f}_2 &&= (\alpha_{21}, \alpha_{22}, 0), \\ \mathbf{e}_3 &= \alpha_{31}\mathbf{f}_1 + \alpha_{32}\mathbf{f}_2 + \alpha_{33}\mathbf{f}_3 &&= (\alpha_{31}, \alpha_{32}, \alpha_{33}). \end{aligned}$$

The coefficient α_{11} is found from the condition

$$A(\mathbf{e}_1; \mathbf{f}_1) = 1,$$

i.e., $2\alpha_{11} = 1$, or $\alpha_{11} = \frac{1}{2}$ and

$$\mathbf{e}_1 = \tfrac{1}{2}\mathbf{f}_1 = (\tfrac{1}{2}, 0, 0).$$

Next α_{21} and α_{22} are determined from the equations

$$A(\mathbf{e}_2; \mathbf{f}_1) = 0 \quad \text{and} \quad A(\mathbf{e}_2, \mathbf{f}_2) = 1,$$

or,

$$2\alpha_{21} + \tfrac{3}{2}\alpha_{22} = 0; \qquad \tfrac{3}{2}\alpha_{21} + \alpha_{22} = 1,$$

whence

$$\alpha_{21} = 6, \qquad \alpha_{22} = -8,$$

and

$$\mathbf{e}_2 = 6\mathbf{f}_1 - 8\mathbf{f}_2 = (6, -8, 0).$$

Finally, α_{31}, α_{32}, α_{33} are determined from the equations

$$A(\mathbf{e}_3; \mathbf{f}_1) = 0, \qquad A(\mathbf{e}_3; \mathbf{f}_2) = 0, \qquad A(\mathbf{e}_3; \mathbf{f}_3) = 1$$

or

$$\begin{aligned} 2\alpha_{31} + \tfrac{3}{2}\alpha_{32} + 2\alpha_{33} &= 0, \\ \tfrac{3}{2}\alpha_{31} + \alpha_{32} \qquad &= 0, \\ 2\alpha_{31} \qquad + \alpha_{33} &= 1, \end{aligned}$$

whence

$$\alpha_{31} = \frac{8}{17}, \qquad \alpha_{32} = -\frac{12}{17}, \qquad \alpha_{33} = \frac{1}{17},$$

and

$$\mathbf{e}_3 = \frac{8}{17}\mathbf{f}_1 - \frac{12}{17}\mathbf{f}_2 + \frac{1}{17}\mathbf{f}_3 = \left(\frac{8}{17}, -\frac{12}{17}, \frac{1}{17}\right).$$

Relative to the basis $\mathbf{e}_1$, $\mathbf{e}_2$, $\mathbf{e}_3$ our quadratic form becomes

$$A(\mathbf{x}; \mathbf{x}) = \frac{1}{\Delta_1}\zeta_1^2 + \frac{\Delta_1}{\Delta_2}\zeta_2^2 + \frac{\Delta_2}{\Delta_3}\zeta_3^2 = \tfrac{1}{2}\zeta_1^2 - 8\zeta_2^2 + \frac{1}{17}\zeta_3^2.$$

Here ζ_1, ζ_2, ζ_3 are the coordinates of the vector $\mathbf{x}$ in the basis $\mathbf{e}_1$, $\mathbf{e}_2$, $\mathbf{e}_3$.

2. In proving Theorem 1 above we not only constructed a basis in which the given quadratic form is expressed as a sum of squares but we also obtained expressions for the coefficients that go with these squares. These coefficients are

$$\frac{1}{\Delta_1}, \quad \frac{\Delta_1}{\Delta_2}, \cdots, \frac{\Delta_{n-1}}{\Delta_n},$$

so that the quadratic form is

$$\frac{1}{\Delta_1}\xi_1^2 + \frac{\Delta_1}{\Delta_2}\xi_2^2 + \cdots + \frac{\Delta_{n-1}}{\Delta_n}\xi_n^2. \tag{8}$$

It is clear that if Δ_{i-1} and Δ_i have the same sign then the coefficient of ξ_i^2 is positive and that if Δ_{i-1} and Δ_i have opposite signs, then this coefficient is negative. Hence,

THEOREM 2. *The number of negative coefficients which appear in the canonical form* (8) *of a quadratic form is equal to the number of changes of sign in the sequence*

$$1, \Delta_1, \Delta_2, \cdots, \Delta_n.$$

Actually, all we have shown is how to compute the number of positive and negative squares for a particular mode of reducing a quadratic form to a sum of squares. In the next section we shall show that the number of positive and negative squares is independent of the method used in reducing the form to a sum of squares.

Assume that $\Delta_1 > 0$, $\Delta_2 > 0, \cdots, \Delta_n > 0$. Then there exists a basis $\mathbf{e}_1, \mathbf{e}_2, \cdots, \mathbf{e}_n$ in which $A(\mathbf{x}; \mathbf{x})$ takes the form

$$A(\mathbf{x}; \mathbf{x}) = \lambda_1\xi_1^2 + \lambda_2\xi_2^2 + \cdots + \lambda_n\xi_n^2,$$

where all the λ_i are positive. Hence $A(\mathbf{x}; \mathbf{x}) \geqq 0$ for all $\mathbf{x}$ and

$$A(\mathbf{x}; \mathbf{x}) = \sum_{i=1}^{n} \lambda_i\xi_i^2 = 0$$

is equivalent to

$$\xi_1 = \xi_2 = \cdots = \xi_n = 0.$$

In other words,

If $\Delta_1 > 0$, $\Delta_2 > 0, \cdots, \Delta_n > 0$, *then the quadratic form* $A(\mathbf{x}; \mathbf{x})$ *is positive definite.*

Conversely, let $A(\mathbf{x}; \mathbf{x})$ be a positive definite quadratic form. We shall show that then

$$\Delta_k > 0 \quad (k = 1, 2, \cdots, n).$$

We first disprove the possibility that

$$\Delta_k = \begin{vmatrix} A(\mathbf{f}_1; \mathbf{f}_1) & A(\mathbf{f}_1; \mathbf{f}_2) & \cdots & A(\mathbf{f}_1; \mathbf{f}_k) \\ A(\mathbf{f}_2; \mathbf{f}_1) & A(\mathbf{f}_2; \mathbf{f}_2) & \cdots & A(\mathbf{f}_2; \mathbf{f}_k) \\ \cdots & \cdots & \cdots & \cdots \\ A(\mathbf{f}_k; \mathbf{f}_1) & A(\mathbf{f}_k; \mathbf{f}_2) & \cdots & A(\mathbf{f}_k; \mathbf{f}_k) \end{vmatrix} = 0.$$

If $\Delta_k = 0$, then one of the rows in the above determinant would be a linear combination of the remaining rows, i.e., it would be possible to find numbers $\mu_1, \mu_2, \cdots, \mu_k$ not all zero such that

$$\mu_1 A(\mathbf{f}_1; \mathbf{f}_i) + \mu_2 A(\mathbf{f}_2; \mathbf{f}_i) + \cdots + \mu_k A(\mathbf{f}_k; \mathbf{f}_i) = 0,$$

$i = 1, 2, \cdots, k$. But then

$$A(\mu_1 \mathbf{f}_1 + \mu_2 \mathbf{f}_2 + \cdots + \mu_k \mathbf{f}_k; \mathbf{f}_i) = 0 \quad (i = 1, 2, \cdots, k),$$

so that

$$A(\mu_1 \mathbf{f}_1 + \mu_2 \mathbf{f}_2 + \cdots + \mu_k \mathbf{f}_k; \mu_1 \mathbf{f}_1 + \mu_2 \mathbf{f}_2 + \cdots + \mu_k \mathbf{f}_k) = 0.$$

In view of the fact that $\mu_1 \mathbf{f}_1 + \mu_2 \mathbf{f}_2 + \cdots + \mu_k \mathbf{f}_k \neq 0$, the latter equality is incompatible with the assumed positive definite nature of our form.

The fact that $\Delta_k \neq 0$ $(\mathbf{k} = 1, \cdots, n)$ combined with Theorem 1 permits us to conclude that it is possible to express $A(\mathbf{x}; \mathbf{x})$ in the form

$$A(\mathbf{x}; \mathbf{x}) = \lambda_1 \xi_1^2 + \lambda_2 \xi_2^2 + \cdots + \lambda_2 \xi_n^2, \quad \lambda_k = \frac{\Delta_{k-1}}{\Delta_k}.$$

Since for a positive definite quadratic form all $\lambda_k > 0$, it follows that all $\Delta_k > 0$ (we recall that $\Delta_0 = 1$).

We have thus proved

THEOREM 3. *Let $A(\mathbf{x}; \mathbf{y})$ be a symmetric bilinear form and $\mathbf{f}_1, \mathbf{f}_2, \cdots, \mathbf{f}_n$, a basis of the n-dimensional space* $\mathbf{R}$. *For the quadratic form $A(\mathbf{x}; \mathbf{x})$ to be positive definite it is necessary and sufficient that*

$$\Delta_1 > 0, \Delta_2 > 0, \cdots, \Delta_n > 0.$$

This theorem is known as the Sylvester criterion for a quadratic form to be positive definite.

It is clear that we could use an arbitrary basis of $\mathbf{R}$ to express the conditions for the positive definiteness of the form $A(\mathbf{x}; \mathbf{x})$. In particular if we used as another basis the vectors $\mathbf{f}_1, \mathbf{f}_2, \cdots, \mathbf{f}_n$ in changed order, then the new $\Delta_1, \Delta_2, \cdots, \Delta_n$ would be different principal minors of the matrix $||a_{ik}||$. This implies the following interesting

COROLLARY. *If the principal minors $\Delta_1, \Delta_2, \cdots, \Delta_n$ of a matrix $||a_{ik}||$ of a quadratic form $A(\mathbf{x}; \mathbf{x})$ relative to some basis are positive, then all principal minors of that matrix are positive.*

Indeed, if $\Delta_1, \Delta_2, \cdots, \Delta_n$ are all positive, then $A(\mathbf{x}; \mathbf{x})$ is positive definite. Now let Δ be a principal minor of $||a_{ik}||$ and let $p_1, p_2, \cdots, p_k$ be the numbers of the rows and columns of $||a_{ik}||$ in Δ. If we permute the original basis vectors so that the p_ith vector occupies the ith position ($i = 1, \cdots, k$) and express the conditions for positive definiteness of $A(\mathbf{x}; \mathbf{x})$ relative to the new basis, we see that $\Delta > 0$.

3. *The Gramm determinant.* The results of this section are valid for quadratic forms $A(\mathbf{x}; \mathbf{x})$ derivable from inner products, i.e., for quadratic forms $A(\mathbf{x}; \mathbf{x})$ such that

$$A(\mathbf{x}; \mathbf{x}) \equiv (\mathbf{x}, \mathbf{x}).$$

If $A(\mathbf{x}; \mathbf{y})$ is a symmetric bilinear form on a vector space $\mathbf{R}$ and $A(\mathbf{x}; \mathbf{x})$ is positive definite, then $A(\mathbf{x}; \mathbf{y})$ can be taken as an inner product in $\mathbf{R}$, i.e., we may put $(\mathbf{x}, \mathbf{y}) \equiv A(\mathbf{x}; \mathbf{y})$. Conversely, if $(\mathbf{x}, \mathbf{y})$ is an inner product on $\mathbf{R}$, then $A(\mathbf{x}; \mathbf{y}) \equiv (\mathbf{x}, \mathbf{y})$ is a bilinear symmetric form on $\mathbf{R}$ such that $A(\mathbf{x}; \mathbf{x})$ is positive definite. Thus every positive definite quadratic form on $\mathbf{R}$ may be identified with an inner product on $\mathbf{R}$ considered for pairs of equal vectors only, $A(\mathbf{x}; \mathbf{x}) \equiv (\mathbf{x}, \mathbf{x})$. One consequence of this correspondence is that every theorem concerning positive definite quadratic forms is at the same time a theorem about vectors in Euclidean space.

Let $\mathbf{e}_1, \mathbf{e}_2, \cdots, \mathbf{e}_k$ be $\mathbf{k}$ vectors in some Euclidean space. The determinant

$$\begin{vmatrix} (\mathbf{e}_1, \mathbf{e}_1) & (\mathbf{e}_1, \mathbf{e}_2) & \cdots & (\mathbf{e}_1, \mathbf{e}_k) \\ (\mathbf{e}_2, \mathbf{e}_1) & (\mathbf{e}_2, \mathbf{e}_2) & \cdots & (\mathbf{e}_2, \mathbf{e}_k) \\ \cdots & \cdots & \cdots & \cdots \\ (\mathbf{e}_k, \mathbf{e}_1) & (\mathbf{e}_k, \mathbf{e}_2) & \cdots & (\mathbf{e}_k, \mathbf{e}_k) \end{vmatrix}$$

is known as the *Gramm determinant* of these vectors.

THEOREM 4. *The Gramm determinant of a system of vectors $\mathbf{e}_1, \mathbf{e}_2, \cdots, \mathbf{e}_k$ is always $\geqq 0$. This determinant is zero if and only if the vectors $\mathbf{e}_1, \mathbf{e}_2, \cdots, \mathbf{e}_k$ are linearly dependent.*

Proof: Assume that $\mathbf{e}_1, \mathbf{e}_2, \cdots, \mathbf{e}_k$ are linearly independent. Consider the bilinear form $A(\mathbf{x}; \mathbf{y}) \equiv (\mathbf{x}, \mathbf{y})$, where $(\mathbf{x}, \mathbf{y})$ is the inner product of $\mathbf{x}$ and $\mathbf{y}$. Then the Gramm determinant of $\mathbf{e}_1, \mathbf{e}_2, \cdots, \mathbf{e}_k$ coincides with the determinant Δ_k discussed in this section (cf. (7)). Since $A(\mathbf{x}; \mathbf{y})$ is a symmetric bilinear form such that $A(\mathbf{x}; \mathbf{x})$ is positive definite it follows from Theorem 3 that $\Delta_k > 0$.

We shall show that the Gramm determinant of a system of linearly dependent vectors $\mathbf{e}_1, \mathbf{e}_2, \cdots, \mathbf{e}_k$ is zero. Indeed, in that case one of the vectors, say $\mathbf{e}_k$, is a linear combination of the others,

$$\mathbf{e}_k = \lambda_1 \mathbf{e}_1 + \lambda_2 \mathbf{e}_2 + \cdots + \lambda_{k-1} \mathbf{e}_{k-1}.$$

It follows that the last row in the Gramm determinant of the vectors $\mathbf{e}_1, \mathbf{e}_2, \cdots, \mathbf{e}_k$ is a linear combination of the others and the determinant must vanish. This completes the proof.

As an example consider the Gramm determinant of two vectors $\mathbf{x}$ and $\mathbf{y}$

$$\Delta_2 = \begin{vmatrix} (\mathbf{x}, \mathbf{x}) & (\mathbf{x}, \mathbf{y}) \\ (\mathbf{y}, \mathbf{x}) & (\mathbf{y}, \mathbf{y}) \end{vmatrix}$$

The assertion that $\Delta_2 > 0$ is synonymous with the Schwarz inequality.

EXAMPLES. *1*. In Euclidean three-space (or in the plane) the determinant Δ_2 has the following geometric sense: Δ_2 is the square of the area of the parallelogram with sides $\mathbf{x}$ and $\mathbf{y}$. Indeed,

$$(\mathbf{x}, \mathbf{y}) = (\mathbf{y}, \mathbf{x}) = |\mathbf{x}| \cdot |\mathbf{y}| \cos \varphi,$$

where φ is the angle between $\mathbf{x}$ and $\mathbf{y}$. Therefore,

$$\Delta_2 = |\mathbf{x}|^2 |\mathbf{y}|^2 - |\mathbf{x}|^2 |\mathbf{y}|^2 \cos^2 \varphi = |\mathbf{x}|^2 |\mathbf{y}|^2 (1 - \cos^2 \varphi) = |\mathbf{x}|^2 |\mathbf{y}|^2 \sin^2 \varphi,$$

i.e., Δ_2 has indeed the asserted geometric meaning.

2. In three-dimensional Euclidean space the volume of a parallelepiped on the vectors $\mathbf{x}$, $\mathbf{y}$, $\mathbf{z}$ is equal to the absolute value of the determinant

$$v = \begin{vmatrix} x_1 & x_2 & x_3 \\ y_1 & y_2 & y_3 \\ z_1 & z_2 & z_3 \end{vmatrix}.$$

where x_i, y_i, z_i are the Cartesian coordinates of $\mathbf{x}$, $\mathbf{y}$, $\mathbf{z}$. Now,

$$v^2 = \begin{vmatrix} x_1^2 + x_2^2 + x_3^2 & x_1y_1 + x_2y_2 + x_3y_3 & x_1z_1 + x_2z_2 + x_3z_3 \\ y_1x_1 + y_2x_2 + y_3x_3 & y_1^2 + y_2^2 + y_3^2 & y_1z_1 + y_2z_2 + y_3z_3 \\ z_1x_1 + z_2x_2 + z_3x_3 & z_1y_1 + z_2y_2 + z_3y_3 & z_1^2 + z_2^2 + z_3^2 \end{vmatrix} =$$

$$= \begin{vmatrix} (\mathbf{x}, \mathbf{x}) & (\mathbf{x}, \mathbf{y}) & (\mathbf{x}, \mathbf{z}) \\ (\mathbf{y}, \mathbf{x}) & (\mathbf{y}, \mathbf{y}) & (\mathbf{y}, \mathbf{z}) \\ (\mathbf{z}, \mathbf{x}) & (\mathbf{z}, \mathbf{y}) & (\mathbf{z}, \mathbf{z}) \end{vmatrix}.$$

Thus the Gramm determinant of three vectors **x**, **y**, **z** is the square of the volume of the parallelepiped on these vectors.

Similarly, it is possible to show that the Gramm determinant of k vectors **x**, **y**, $\cdots$, **w** in a k-dimenional space **R** is the square of the determinant

$$\begin{vmatrix} x_1 & x_2 & \cdots & x_k \\ y_1 & y_2 & \cdots & y_k \\ \cdots & \cdots & \cdots & \cdots \\ w_1 & w_2 & \cdots & w_k \end{vmatrix}, \tag{9}$$

where the x_i are coordinates of **x** in some orthogonal basis, the y_i are the coordinates of **y** in that basis, etc.

(It is clear that the space **R** need not be k-dimensional. **R** may, indeed, be even infinite-dimensional since our considerations involve only the subspace generated by the k vectors **x**, **y**, $\cdots$, **w**.)

By analogy with the three-dimensional case, the determinant (9) is referred to as the volume of the k-dimensional parallelepiped determined by the vectors **x**, **y**, $\cdots$, **w**.

3. In the space of functions (Example *4*, § 2) the Gramm determinant takes the form

$$\Delta = \begin{vmatrix} \int_a^b f_1^2(t)dt & \int_a^b f_1(t)f_2(t)dt & \cdots & \int_a^b f_1(t)f_k(t)dt \\ \int_a^b f_2(t)f_1(t)dt & \int_a^b f_2^2(t)dt & \cdots & \int_a^b f_2(t)f_k(t)dt \\ \cdots & \cdots & \cdots & \cdots \\ \int_a^b f_k(t)f_1(t)dt & \int_a^b f_k(t)f_2(t)dt & \cdots & \int_a^b f_k^2(t)dt \end{vmatrix},$$

and the theorem just proved implies that:

The Gramm determinant of a system of functions is always $\geqq 0$. *For a system of functions to be linearly dependent it is necessary and sufficient that their Gramm determinant vanish.*

§ 7. The law of inertia

1. *The law of inertia.* There are different bases relative to which a quadratic form $A(\mathbf{x}; \mathbf{x})$ is a sum of squares,

$$A(\mathbf{x}; \mathbf{x}) = \sum_{i=1}^{n} \lambda_i \xi_i^2 \tag{1}$$

By replacing those basis vectors (in such a basis) which correspond to the non-zero λ_i by vectors proportional to them we obtain a representation of $A(\mathbf{x}; \mathbf{x})$ by means of a sum of squares in which the λ_i are 0, 1, or -1. It is natural to ask whether the number of coefficients whose values are respectively 0, 1, and -1 is dependent on the choice of basis or is solely dependent on the quadratic form $A(\mathbf{x}; \mathbf{x})$.

To illustrate the nature of the question consider a quadratic form $A(\mathbf{x}; \mathbf{x})$ which, relative to some basis $\mathbf{e}_1, \mathbf{e}_2, \cdots, \mathbf{e}_n$, is represented by the matrix

$$||a_{ik}||,$$

where $a_{ik} = A(\mathbf{e}_i; \mathbf{e}_k)$ and all the determinants

$$\Delta_1 = a_{11}, \qquad \Delta_2 = \begin{vmatrix} a_{11} & a_{12} \\ a_{21} & a_{22} \end{vmatrix}, \qquad \cdots,$$

$$\Delta_n = \begin{vmatrix} a_{11} & a_{12} & \cdots & a_{1n} \\ a_{21} & a_{22} & \cdots & a_{2n} \\ \cdots & \cdots & \cdots & \cdots \\ a_{n1} & a_{n2} & \cdots & a_{nn} \end{vmatrix}$$

are different from zero. Then, as was shown in para. 2, § 6, all λ_i in formula (1) are different from zero and the number of positive coefficients obtained after reduction of $A(\mathbf{x}; \mathbf{x})$ to a sum of squares *by the method described in that section* is equal to the number of changes of sign in the sequence $1, \Delta_1, \Delta_2, \cdots, \Delta_n$.

Now, suppose some other basis $\mathbf{e}'_1, \mathbf{e}'_2, \cdots, \mathbf{e}'_n$ were chosen. Then a certain matrix $||a'_{ik}||$ would take the place of $||a_{ik}||$ and certain determinants

$$\Delta'_1, \Delta'_2, \cdots, \Delta'_n$$

would replace the determinants $\Delta_1, \Delta_2, \cdots, \Delta_n$. There arises the question of the connection (if any) between the number of changes of sign in the squences $1, \Delta'_1, \Delta'_2, \cdots, \Delta'_n$ and $1, \Delta_1, \Delta_2, \cdots, \Delta_n$.

The following theorem, known as *the law of inertia of quadratic forms*, answers the question just raised.

THEOREM 1. *If a quadratic form is reduced by two different methods (i.e., in two different bases) to a sum of squares, then the number of positive coefficients as well as the number of negative coefficients is the same in both cases.*

Theorem 1 states that the number of positive λ_i in (1) and the number of negative λ_i in (1) are invariants of the quadratic form. Since the total number of the λ_i is n, it follows that *the number of coefficients λ_i which vanish is also an invariant of the form.*

We first prove the following lemma:

LEMMA. *Let $\mathbf{R}'$ and $\mathbf{R}''$ be two subspaces of an n-dimensional space $\mathbf{R}$ of dimension k and l, respectively, and let $k + l > n$. Then there exists a vector $\mathbf{x} \neq \mathbf{0}$ contained in $\mathbf{R}' \cap \mathbf{R}''$.*

Proof: Let $\mathbf{e}_1, \mathbf{e}_2, \cdots, \mathbf{e}_k$ be a basis of $\mathbf{R}'$ and $\mathbf{f}_1, \mathbf{f}_2, \cdots, \mathbf{f}_l$, basis of $\mathbf{R}''$. The vectors $\mathbf{e}_1, \mathbf{e}_2, \cdots, \mathbf{e}_k, \mathbf{f}_1, \mathbf{f}_2, \cdots, \mathbf{f}_l$ are linearly dependent $(k + l > n)$. This means that there exist numbers $\lambda_1, \lambda_2, \cdots, \lambda_k, \mu_1, \mu_2, \cdots, \mu_l$ not all zero such that

$$\lambda_1\mathbf{e}_1 + \lambda_2\mathbf{e}_2 + \cdots + \lambda_k\mathbf{e}_k + \mu_1\mathbf{f}_1 + \mu_2\mathbf{f}_2 + \cdots + \mu_l\mathbf{f}_l = \mathbf{0},$$

i.e.,

$$\lambda_1\mathbf{e}_1 + \lambda_2\mathbf{e}_2 + \cdots + \lambda_k\mathbf{e}_k = -\mu_1\mathbf{f}_1 - \mu_2\mathbf{f}_2 - \cdots - \mu_l\mathbf{f}_l.$$

Let us put

$$\lambda_1\mathbf{e}_1 + \lambda_2\mathbf{e}_2 + \cdots + \lambda_k\mathbf{e}_k = -\mu_1\mathbf{f}_1 - \mu_2\mathbf{f}_2 - \cdots - \mu_l\mathbf{f}_l = \mathbf{x}.$$

It is clear that $\mathbf{x}$ is in $\mathbf{R}' \cap \mathbf{R}''$. It remains to show that $\mathbf{x} \neq \mathbf{0}$. If $\mathbf{x} = \mathbf{0}$, $\lambda_1, \lambda_2, \cdots, \lambda_k$ and $\mu_1, \mu_2, \cdots, \mu_l$ would all be zero, which is impossible. Hence $\mathbf{x} \neq \mathbf{0}$.

We can now prove Theorem 1.

Proof: Let $\mathbf{e}_1, \mathbf{e}_2, \cdots, \mathbf{e}_n$ be a basis in which the quadratic form $A(\mathbf{x}; \mathbf{x})$ becomes

$$(2)\quad A(\mathbf{x}; \mathbf{x}) = \xi_1^2 + \xi_2^2 + \cdots + \xi_p^2 - \xi^2_{p+1} - \xi^2_{p+2} - \cdots - \xi^2_{p+q}.$$

(Here $\xi_1, \xi_2, \cdots, \xi_n$ are the coordinates of the vector $\mathbf{x}$, i.e., $\mathbf{x} = \xi_1\mathbf{e}_1 + \xi_2\mathbf{e}_2 + \cdots + \xi_p\mathbf{e}_p + \xi_{p+1}\mathbf{e}_{p+1} + \cdots + \xi_{p+q}\mathbf{e}_{p+q} + \cdots + \xi_n\mathbf{e}_n$.) Let $\mathbf{f}_1, \mathbf{f}_2, \cdots, \mathbf{f}_n$ be another basis relative to which the quadratic form becomes

$$(3)\quad A(\mathbf{x}; \mathbf{x}) = \eta_1^2 + \eta_2^2 + \cdots + \eta^2_{p'} - \eta^2_{p'+1} - \cdots - \eta^2_{p'+q'}.$$

(Here $\eta_1, \eta_2, \cdots, \eta_n$ are the coordinates of $\mathbf{x}$ relative to the basis $\mathbf{f}_1, \mathbf{f}_2, \cdots, \mathbf{f}_n$.) We must show that $p = p'$ and $q = q'$. Assume that this is false and that $p > p'$, say.

Let $\mathbf{R}'$ be the subspace spanned by the vectors $\mathbf{e}_1, \mathbf{e}_2, \cdots, \mathbf{e}_p$.

$\mathbf{R}'$ has dimension p. The subspace $\mathbf{R}''$ spanned by the vectors $\mathbf{f}_{p'+1}, \mathbf{f}_{p'+2}, \cdots, \mathbf{f}_n$ has dimension $n - p'$. Since $n - p' + p > n$ (we assumed $p > p'$), there exists a vector $\mathbf{x} \neq \mathbf{0}$ in $\mathbf{R}' \cap \mathbf{R}''$ (cf. Lemma), i.e.,

$$\mathbf{x} = \xi_1 \mathbf{e}_1 + \xi_2 \mathbf{e}_2 + \cdots + \xi_p \mathbf{e}_p$$

and

$$\mathbf{x} = \eta_{p'+1} \mathbf{f}_{p'+1} + \cdots + \eta_{p'+q'} \mathbf{f}_{p'+q'} + \cdots + \eta_n \mathbf{f}_n.$$

The coordinates of the vector $\mathbf{x}$ relative to the basis $\mathbf{e}_1, \mathbf{e}_2, \cdots, \mathbf{e}_n$ are $\xi_1, \xi_2, \cdots, \xi_p, 0, \cdots, 0$ and its coordinates relative to the basis $\mathbf{f}_1, \mathbf{f}_2, \cdots, \mathbf{f}_n$ are $0, 0, \cdots, 0, \eta_{p'+1}, \cdots, \eta_n$. Substituting these coordinates in (2) and (3) respectively we get, on the one hand,

$$A(\mathbf{x}; \mathbf{x}) = \xi_1^2 + \xi_2^2 + \cdots + \xi_p^2 > 0 \tag{4}$$

(since not all the ξ_i vanish) and, on the other hand,

$$A(\mathbf{x}; \mathbf{x}) = -\eta^2_{p'+1} - \eta^2_{p'+2} - \cdots - \eta^2_{p'+q'} \leqq 0. \tag{5}$$

(Note that it is not possible to replace $\leqq$ in (5) with $<$, for, while not all the numbers $\eta_{p'+1}, \cdots, \eta_n$ are zero, it is possible that $\eta_{p'+1} = \eta_{p'+2} = \cdots = \eta_{p'+q'} = 0$.) The resulting contradiction shows that $p = p'$. Similarly one can show that $q = q'$. This completes the proof of the law of inertia of quadratic forms.

2. *Rank of a quadratic form*

DEFINITION 1. *By the rank of a quadratic form we mean the number of non-zero coefficients λ_i in one of its canonical forms.*

The reasonableness of the above definition follows from the law of inertia just proved. We shall now investigate the problem of actually finding the rank of a quadratic form. To this end we shall define the rank of a quadratic form without recourse to its canonical form.

DEFINITION 2. *By the null space of a given bilinear form $A(\mathbf{x}; \mathbf{y})$ we mean the set R_0 of all vectors $\mathbf{y}$ such that $A(\mathbf{x}; \mathbf{y}) = 0$ for every $\mathbf{x} \in \mathbf{R}$.*

It is easy to see that $\mathbf{R}_0$ is a subspace of $\mathbf{R}$. Indeed, let $\mathbf{y}_1, \mathbf{y}_2 \in \mathbf{R}_0$, i.e., $A(\mathbf{x}; \mathbf{y}_1) = 0$ and $A(\mathbf{x}; \mathbf{y}_2) = 0$ for all $\mathbf{x} \in \mathbf{R}$. Then $A(\mathbf{x}; \mathbf{y}_1 + \mathbf{y}_2) = 0$ and $A(\mathbf{x}; \lambda\mathbf{y}_1) = 0$ for all $\mathbf{x} \in \mathbf{R}$. But this means that $\mathbf{y}_1 + \mathbf{y}_2 \in \mathbf{R}_0$ and $\lambda\mathbf{y}_1 \in \mathbf{R}_0$.

We shall now try to get a better insight into the space $\mathbf{R}_0$. If $\mathbf{f}_1, \mathbf{f}_2, \cdots, \mathbf{f}_n$ is a basis of $\mathbf{R}$, then for a vector

$$\mathbf{y} = \eta_1\mathbf{f}_1 + \eta_2\mathbf{f}_2 + \cdots + \eta_n\mathbf{f}_n \tag{6}$$

to belong to the null space of $A(\mathbf{x}; \mathbf{y})$ it suffices that

$$A(\mathbf{f}_i; \mathbf{y}) = 0 \quad \text{for } i = 1, 2, \cdots, n. \tag{7}$$

Replacing $\mathbf{y}$ in (7) by (6) we obtain the following system of equations:

$$\begin{aligned} A(\mathbf{f}_1; \eta_1\mathbf{f}_1 + \eta_2\mathbf{f}_2 + \cdots + \eta_n\mathbf{f}_n) &= 0, \\ A(\mathbf{f}_2; \eta_1\mathbf{f}_1 + \eta_2\mathbf{f}_2 + \cdots + \eta_n\mathbf{f}_n) &= 0, \\ \cdots\cdots\cdots\cdots\cdots\cdots\cdots\cdots & \\ A(\mathbf{f}_n; \eta_1\mathbf{f}_1 + \eta_2\mathbf{f}_2 + \cdots + \eta_n\mathbf{f}_n) &= 0. \end{aligned}$$

If we put $A(\mathbf{f}_i; \mathbf{f}_k) = a_{ik}$, the above system goes over into

$$\begin{aligned} a_{11}\eta_1 + a_{12}\eta_2 + \cdots + a_{1n}\eta_n &= 0, \\ a_{21}\eta_1 + a_{22}\eta_2 + \cdots + a_{2n}\eta_n &= 0, \\ \cdots\cdots\cdots\cdots\cdots\cdots\cdots\cdots & \\ a_{n1}\eta_1 + a_{n2}\eta_2 + \cdots + a_{nn}\eta_n &= 0. \end{aligned}$$

Thus the null space $\mathbf{R}_0$ consists of all vectors $\mathbf{y}$ whose coordinates $\eta_1, \eta_2, \cdots, \eta_n$ are solutions of the above system of linear equations. As is well known, the dimension of this subspace is $n - r$, where r is the rank of the matrix $||a_{ik}||$.

We can now argue that

The rank of the matrix $||a_{ik}||$ of the bilinear form $A(\mathbf{x}; \mathbf{y})$ is independent of the choice of basis in $\mathbf{R}$ (although the matrix $||a_{ik}||$ does depend on the choice of basis; cf. § 5).

Indeed, the rank of the matrix in question is $n - r_0$, where r_0 is the dimension of the null space, and the null space is completely independent of the choice of basis.

We shall now connect the rank of the matrix of a quadratic form with the rank of the quadratic form. We defined the rank of a quadratic form to be the number of (non-zero) squares in any of its canonical forms. But relative to a canonical basis the matrix of a quadratic form is diagonal

$$\begin{bmatrix} \lambda_1 & 0 & \cdots & 0 \\ 0 & \lambda_2 & \cdots & 0 \\ \cdots & \cdots & \cdots & \cdots \\ 0 & 0 & \cdots & \lambda_n \end{bmatrix}$$

and its rank r is equal to the number of non-zero coefficients, i.e., the rank of the quadratic form. Since we have shown that the rank of the matrix of a quadratic form does not depend on the choice of basis, the rank of the matrix associated with a quadratic form in any basis is the same as the rank of the quadratic form. [5]

To sum up:

THEOREM 2. *The matrices which represent a quadratic form in different coordinate systems all have the same rank r. This rank is equal to the number of squares with non-zero multipliers in any canonical form of the quadratic form.*

Thus, to find the rank of a quadratic form we must compute the rank of its matrix relative to an arbitrary basis.

§ 8. Complex n-dimensional space

In the preceding sections we dealt essentially with vector spaces over the field of real numbers. Many of the results presented so far remain in force for vector spaces over arbitrary fields. In addition to vector spaces over the field of real numbers, vector spaces over the field of complex numbers will play a particularly important role in the sequel. It is therefore reasonable to discuss the contents of the preceding sections with this case in mind.

1. *Complex vector spaces.* We mentioned in § 1 that all of the results presented in that section apply to vector spaces over arbitrary fields and, in particular, to vector spaces over the field of complex numbers.

2. *Complex Euclidean vector spaces.* By a complex Euclidean vector space we mean a complex vector space in which there is defined an inner product, i.e., a function which associates with every pair of vectors $\mathbf{x}$ and $\mathbf{y}$ a complex number $(\mathbf{x}, \mathbf{y})$ so that the following axioms hold:

1. $(\mathbf{x}, \mathbf{y}) = \overline{(\mathbf{y}, \mathbf{x})}$ [$\overline{(\mathbf{y}, \mathbf{x})}$ denotes the complex conjugate of $(\mathbf{y}, \mathbf{x})$];

[5] We could have obtained the same result by making use of the well-known fact that the rank of a matrix is not changed if we multiply it by any non-singular matrix and by noting that the connection between two matrices $\mathscr{A}$ and $\mathscr{B}$ which represent the same quadratic form relative to two different bases is $\mathscr{B} = \mathscr{C}' \mathscr{A} \mathscr{C}$, $\mathscr{C}$ non-singular.

2. $(\lambda \mathbf{x}, \mathbf{y}) = \lambda(\mathbf{x}, \mathbf{y})$;
3. $(\mathbf{x}_1 + \mathbf{x}_2, \mathbf{y}) = (\mathbf{x}_1, \mathbf{y}) + (\mathbf{x}_2, \mathbf{y})$;
4. $(\mathbf{x}, \mathbf{x})$ is a non-negative real number which becomes zero only if $\mathbf{x} = \mathbf{0}$.

Complex Euclidean vector spaces are referred to as *unitary spaces*.

Axioms 1 and 2 imply that $(\mathbf{x}, \lambda\mathbf{y}) = \bar{\lambda}(\mathbf{x}, \mathbf{y})$. In fact,

$$(\mathbf{x}, \lambda\mathbf{y}) = \overline{(\lambda\mathbf{y}, \mathbf{x})} = \bar{\lambda}\overline{(\mathbf{y}, \mathbf{x})} = \bar{\lambda}(\mathbf{x}, \mathbf{y}).$$

Also, $(\mathbf{x}, \mathbf{y}_1 + \mathbf{y}_2) = (\mathbf{x}, \mathbf{y}_1) + (\mathbf{x}, \mathbf{y}_2)$. Indeed,

$$(\mathbf{x}, \mathbf{y}_1 + \mathbf{y}_2) = \overline{(\mathbf{y}_1 + \mathbf{y}_2, \mathbf{x})} = \overline{(\mathbf{y}_1, \mathbf{x})} + \overline{(\mathbf{y}_2, \mathbf{x})} = (\mathbf{x}, \mathbf{y}_1) + (\mathbf{x}, \mathbf{y}_2).$$

Axiom 1 above differs from the corresponding Axiom 1 for a real Euclidean vector space. This is justified by the fact that in unitary spaces it is not possible to retain Axioms 1, 2 and 4 for inner products in the form in which they are stated for real Euclidean vector spaces. Indeed,

$$(\mathbf{x}, \mathbf{y}) = (\mathbf{y}, \mathbf{x}).$$

would imply

$$(\mathbf{x}, \lambda\mathbf{y}) = \lambda(\mathbf{x}, \mathbf{y}).$$

But then

$$(\lambda\mathbf{x}, \lambda\mathbf{x}) = \lambda^2(\mathbf{x}, \mathbf{x}).$$

In particular,

$$(i\mathbf{x}, i\mathbf{x}) = -(\mathbf{x}, \mathbf{x}),$$

i.e., the numbers $(\mathbf{x}, \mathbf{x})$ and $(\mathbf{y}, \mathbf{y})$ with $\mathbf{y} = i\mathbf{x}$ would have different signs thus violating Axiom 4.

EXAMPLES OF UNITARY SPACES. *1.* Let $\mathbf{R}$ be the set of n-tuples of complex numbers with the usual definitions of addition and multiplications by (complex) numbers. If

$$\mathbf{x} = (\xi_1, \xi_2, \cdots, \xi_n) \quad \text{and} \quad \mathbf{y} = (\eta_1, \eta_2, \cdots, \eta_n)$$

are two elements of $\mathbf{R}$, we define

$$(\mathbf{x}, \mathbf{y}) = \xi_1\bar{\eta}_1 + \xi_2\bar{\eta}_2 + \cdots + \xi_n\bar{\eta}_n.$$

We leave to the reader the verification of the fact that with the above definition of inner product $\mathbf{R}$ becomes a unitary space.

2. The set $\mathbf{R}$ of Example *1* above can be made into a unitary space by putting

$$(\mathbf{x}, \mathbf{y}) = \sum_{i,k=1}^{n} a_{ik}\xi_i\bar{\eta}_k,$$

where a_{ik} are given complex numbers satisfying the following two conditions:

(α) $a_{ik} = \overline{a_{ki}}$

(β) $\sum a_{ik}\xi_i\bar{\xi}_k \geqq 0$ for every n-tuple $\xi_1, \xi_2, \cdots, \xi_n$ and takes on the value zero only if $\xi_1 = \xi_2 = \cdots = \xi_n = 0$.

3. Let **R** be the set of complex valued functions of a real variable t defined and integrable on an interval $[a, b]$. It is easy to see that **R** becomes a unitary space if we put

$$(f(t), g(t)) = \int_a^b f(t)\overline{g(t)}\, dt.$$

By the *length of a vector* **x** in a unitary space we shall mean the number $\sqrt{(\mathbf{x}, \mathbf{x})}$. Axiom 4 implies that the length of a vector is non-negative and is equal to zero only if the vector is the zero vector.

Two vectors **x** and **y** are said to be *orthogonal* if $(\mathbf{x}, \mathbf{y}) = 0$.

Since the inner product of two vectors is, in general, not a real number, we do not introduce the concept of angle between two vectors.

3. *Orthogonal basis. Isomorphism of unitary spaces.* By an orthogonal basis in an n-dimensional unitary space we mean a set of n pairwise orthogonal non-zero vectors $\mathbf{e}_1, \mathbf{e}_2, \cdots, \mathbf{e}_n$. As in § 3 we prove that the vectors $\mathbf{e}_1, \mathbf{e}_2, \cdots, \mathbf{e}_n$ are linearly independent, i.e., that they form a basis.

The existence of an orthogonal basis in an n-dimensional unitary space is demonstrated by means of a procedure analogous to the orthogonalization procedure described in § 3.

If $\mathbf{e}_1, \mathbf{e}_2, \cdots, \mathbf{e}_n$ is an orthonormal basis and

$$\mathbf{x} = \xi_1\mathbf{e}_1 + \xi_2\mathbf{e}_2 + \cdots + \xi_n\mathbf{e}_n, \quad \mathbf{y} = \eta_1\mathbf{e}_1 + \eta_2\mathbf{e}_2 + \cdots + \eta_n\mathbf{e}_n$$

are two vectors, then

$$\begin{aligned}(\mathbf{x}, \mathbf{y}) &= (\xi_1\mathbf{e}_1 + \xi_2\mathbf{e}_2 + \cdots + \xi_n\mathbf{e}_n, \eta_1\mathbf{e}_1 + \eta_2\mathbf{e}_2 + \cdots + \eta_n\mathbf{e}_n) \\ &= \xi_1\bar{\eta}_1 + \xi_2\bar{\eta}_2 + \cdots + \xi_n\bar{\eta}_n\end{aligned}$$

(cf. Example *1* in this section).

If $\mathbf{e}_1, \mathbf{e}_2, \cdots, \mathbf{e}_n$ is an orthonormal basis and

$$\mathbf{x} = \xi_1\mathbf{e}_1 + \xi_2\mathbf{e}_2 + \cdots + \xi_n\mathbf{e}_n,$$

then

$$(\mathbf{x}, \mathbf{e}_i) = (\xi_1\mathbf{e}_1 + \xi_2\mathbf{e}_2 + \cdots + \xi_n\mathbf{e}_n, \mathbf{e}_i) = \xi_1(\mathbf{e}_1, \mathbf{e}_i) + \xi_2(\mathbf{e}_2, \mathbf{e}_i) + \cdots + \xi_n(\mathbf{e}_n, \mathbf{e}_i),$$

so that

$$(\mathbf{x}, \mathbf{e}_i) = \xi_i.$$

Using the method of § 3 we prove that all unitary spaces of dimension n are isomorphic.

4. *Bilinear and quadratic forms.* With the exception of positive definiteness all the concepts introduced in § 4 retain meaning for vector spaces over arbitrary fields and in particular for complex vector spaces. However, in the case of complex vector spaces there is another and for us more important way of introducing these concepts.

Linear functions of the first and second kind. A complex valued function f defined on a complex space is said to be *a linear function of the first kind* if

1. $f(\mathbf{x} + \mathbf{y}) = f(\mathbf{x}) + f(\mathbf{y})$,
2. $f(\lambda\mathbf{x}) = \lambda f(\mathbf{x})$,

and *a linear function of the second kind* if

1. $f(\mathbf{x} + \mathbf{y}) = f(\mathbf{x}) + f(\mathbf{y})$,
2. $f(\lambda\mathbf{x}) = \bar{\lambda} f(\mathbf{x})$.

Using the method of § 4 one can prove that every linear function of the first kind can be written in the form

$$f(\mathbf{x}) = a_1\xi_1 + a_2\xi_2 + \cdots + a_n\xi_n,$$

where ξ_i are the coordinates of the vector $\mathbf{x}$ relative to the basis $\mathbf{e}_1, \mathbf{e}_2, \cdots, \mathbf{e}_n$ and a_i are constants, $a_i = f(\mathbf{e}_i)$, and that every linear function of the second kind can be written in the form

$$f(\mathbf{x}) = b_1\bar{\xi}_1 + b_2\bar{\xi}_2 + \cdots + b_n\bar{\xi}_n.$$

DEFINITION 1. *We shall say that $A(\mathbf{x}; \mathbf{y})$ is a bilinear form (function) of the vectors $\mathbf{x}$ and $\mathbf{y}$ if:*

1. *for any fixed $\mathbf{y}$, $A(\mathbf{x}; \mathbf{y})$ is a linear function of the first kind of $\mathbf{x}$,*
2. *for any fixed $\mathbf{x}$, $A(\mathbf{x}; \mathbf{y})$ is a linear function of the second kind of $\mathbf{y}$.* In other words,

1. $A(\mathbf{x}_1 + \mathbf{x}_2; \mathbf{y}) = A(\mathbf{x}_1; \mathbf{y}) + A(\mathbf{x}_2; \mathbf{y})$,
   $A(\lambda\mathbf{x}; \mathbf{y}) = \lambda A(\mathbf{x}; \mathbf{y})$,

2. $A(\mathbf{x}; \mathbf{y}_1 + \mathbf{y}_2) = A(\mathbf{x}; \mathbf{y}_1) + A(\mathbf{x}; \mathbf{y}_2)$,
$A(\mathbf{x}; \lambda\mathbf{y}) = \bar{\lambda} A(\mathbf{x}; \mathbf{y})$.

One example of a bilinear form is the inner product in a unitary space

$$A(\mathbf{x}; \mathbf{y}) = (\mathbf{x}, \mathbf{y})$$

considered as a function of the vectors $\mathbf{x}$ and $\mathbf{y}$. Another example is the expression

$$A(\mathbf{x}; \mathbf{y}) = \sum_{i,k=1}^{n} a_{ik} \xi_i \bar{\eta}_k$$

viewed as a function of the vectors

$$\mathbf{x} = \xi_1 \mathbf{e}_1 + \xi_2 \mathbf{e}_2 + \cdots + \xi_n \mathbf{e}_n,$$
$$\mathbf{y} = \eta_1 \mathbf{e}_1 + \eta_2 \mathbf{e}_2 + \cdots + \eta_n \mathbf{e}_n.$$

Let $\mathbf{e}_1, \mathbf{e}_2, \cdots, \mathbf{e}_n$ be a basis of an n-dimensional complex space. Let $A(\mathbf{x}; \mathbf{y})$ be a bilinear form. If $\mathbf{x}$ and $\mathbf{y}$ have the representations

$$\mathbf{x} = \xi_1 \mathbf{e}_1 + \xi_2 \mathbf{e}_2 + \cdots + \xi_n \mathbf{e}_n, \mathbf{y} = \eta_1 \mathbf{e}_1 + \eta_2 \mathbf{e}_2 + \cdots + \eta_n \mathbf{e}_n,$$

then

$$A(\mathbf{x}; \mathbf{y}) = A(\xi_1 \mathbf{e}_1 + \xi_2 \mathbf{e}_2 + \cdots \xi_n \mathbf{e}_n; \eta_1 \mathbf{e}_1 + \eta_2 \mathbf{e}_2 + \cdots + \eta_n \mathbf{e}_n)$$
$$= \sum_{i,k=1}^{n} \xi_i \bar{\eta}_k A(\mathbf{e}_i; \mathbf{e}_k).$$

The matrix $||a_{ik}||$ with

$$a_{ik} = A(\mathbf{e}_i; \mathbf{e}_k)$$

is called the *matrix of the bilinear form* $A(\mathbf{x}; \mathbf{y})$ *relative* to the basis $\mathbf{e}_1, \mathbf{e}_2, \cdots, \mathbf{e}_n$.

If we put $\mathbf{y} = \mathbf{x}$ in a bilinear form $A(\mathbf{x}; \mathbf{y})$ we obtain a function $A(\mathbf{x}; \mathbf{x})$ called a *quadratic* form (in complex space). The connection between bilinear and quadratic forms in complex space is summed up in the following theorem:

Every bilinear form is uniquely determined by its quadratic form. [6]

[6] We recall that in the case of real vector spaces an analogous statement holds only for symmetric bilinear forms (cf. § 4).

Proof: Let $A(\mathbf{x}; \mathbf{x})$ be a quadratic form and let $\mathbf{x}$ and $\mathbf{y}$ be two arbitrary vectors. The four identities[7]:

(I) $A(\mathbf{x}+\mathbf{y}; \mathbf{x}+\mathbf{y}) = A(\mathbf{x}; \mathbf{x}) + A(\mathbf{y}; \mathbf{x}) + A(\mathbf{x}; \mathbf{y}) + A(\mathbf{y}; \mathbf{y})$,

(II) $A(\mathbf{x}+i\mathbf{y}; \mathbf{x}+i\mathbf{y}) = A(\mathbf{x}; \mathbf{x}) + iA(\mathbf{y}; \mathbf{x}) - iA(\mathbf{x}; \mathbf{y}) + A(\mathbf{y}; \mathbf{y})$,

(III) $A(\mathbf{x}-\mathbf{y}; \mathbf{x}-\mathbf{y}) = A(\mathbf{x}; \mathbf{x}) - A(\mathbf{y}; \mathbf{x}) - A(\mathbf{x}; \mathbf{y}) + A(\mathbf{y}; \mathbf{y})$,

(IV) $A(\mathbf{x}-i\mathbf{y}; \mathbf{x}-i\mathbf{y}) = A(\mathbf{x}; \mathbf{x}) - iA(\mathbf{y}; \mathbf{x}) + iA(\mathbf{x}; \mathbf{y}) + A(\mathbf{y}; \mathbf{y})$,

enable us to compute $A(\mathbf{x}; \mathbf{y})$. Namely, if we multiply the equations (I), (II), (III), (IV) by 1, i, -1, $-i$, respectively, and add the results it follows easily that

$$(1)\quad A(\mathbf{x}; \mathbf{y}) = \tfrac{1}{4}\{A(\mathbf{x}+\mathbf{y}; \mathbf{x}+\mathbf{y}) + iA(\mathbf{x}+i\mathbf{y}; \mathbf{x}+i\mathbf{y}) - A(\mathbf{x}-\mathbf{y}; \mathbf{x}-\mathbf{y}) - iA(\mathbf{x}-i\mathbf{y}; \mathbf{x}-i\mathbf{y})\}.$$

Since the right side of (1) involves only the values of the quadratic form associated with the bilinear form under consideration our assertion is proved.

If we multiply equations (I), (II), (III), (IV) by 1, $-i$, -1, i, respectivly, we obtain similarly,

$$(2)\quad A(\mathbf{y}; \mathbf{x}) = \tfrac{1}{4}\{A(\mathbf{x}+\mathbf{y}; \mathbf{x}+\mathbf{y}) - iA(\mathbf{x}+i\mathbf{y}; \mathbf{x}+i\mathbf{y}) - A(\mathbf{x}-\mathbf{y}; \mathbf{x}-\mathbf{y}) + iA(\mathbf{x}-i\mathbf{y}; \mathbf{x}-i\mathbf{y})\}.$$

DEFINITION 2. *A bilinear form is called Hermitian if*

$$A(\mathbf{x}; \mathbf{y}) = \overline{A(\mathbf{y}; \mathbf{x})}.$$

This concept is the analog of a symmetric bilinear form in a real Euclidean vector space.

For a form to be Hermitian it is necessary and sufficient that its matrix $\|a_{ik}\|$ relative to some basis satisfy the condition

$$a_{ik} = \bar{a}_{ki}.$$

Indeed, if the form $A(\mathbf{x}; \mathbf{y})$ is Hermitian, then

$$a_{ik} = A(\mathbf{e}_i; \mathbf{e}_k) = \overline{A(\mathbf{e}_k; \mathbf{e}_i)} = \bar{a}_{ki}.$$

Conversely, if $a_{ik} = \bar{a}_{ki}$, then

$$A(\mathbf{x}; \mathbf{y}) = \sum a_{ik}\xi_i\bar{\eta}_k = \overline{\sum a_{ki}\eta_k\bar{\xi}_i} = \overline{A(\mathbf{y}; \mathbf{x})}.$$

NOTE: If the matrix of a bilinear form satisfies the condition

[7] Note that $A(\mathbf{x}; \lambda\mathbf{y}) = \bar{\lambda}A(\mathbf{x}; \mathbf{y})$, so that, in particular, $A(\mathbf{x}; i\mathbf{y}) = -iA(\mathbf{x}; \mathbf{y})$.

$a_{ik} = \bar{a}_{ki}$, then the same must be true for the matrix of this form relative to any other basis. Indeed, $a_{ik} = \bar{a}_{ki}$ relative to some basis implies that $A(\mathbf{x}; \mathbf{y})$ is a Hermitian bilinear form; but then $a_{ik} = \bar{a}_{ki}$ relative to any other basis.

If a bilinear form is Hermitian, then the associated quadratic form is also called Hermitian. The following result holds:

For a bilinear form $A(\mathbf{x}; \mathbf{y})$ to be Hermitian it is necessary and sufficient that $A(\mathbf{x}; \mathbf{x})$ be real for every vector $\mathbf{x}$.

Proof: Let the form $A(\mathbf{x}; \mathbf{y})$ be Hermitian; i.e., let $A(\mathbf{x}; \mathbf{y}) = \overline{A(\mathbf{y}; \mathbf{x})}$. Then $A(\mathbf{x}; \mathbf{x}) = \overline{A(\mathbf{x}; \mathbf{x})}$, so that the number $A(\mathbf{x}; \mathbf{x})$ is real. Conversely, if $A(\mathbf{x}; \mathbf{x})$ is real for al $\mathbf{x}$, then, in particular, $A(\mathbf{x} + \mathbf{y}; \mathbf{x} + \mathbf{y})$, $A(\mathbf{x} + i\mathbf{y}; \mathbf{x} + i\mathbf{y})$, $A(\mathbf{x} - \mathbf{y}; \mathbf{x} - \mathbf{y})$, $A(\mathbf{x} - i\mathbf{y}; \mathbf{x} - i\mathbf{y})$ are all real and it is easy to see from formulas (1) and (2) that $A(\mathbf{x}; \mathbf{y}) = \overline{A(\mathbf{y}; \mathbf{x})}$.

COROLLARY. *A quadratic form is Hermitian if and only if it is real valued.*

The proof is a direct consequence of the fact just proved that for a bilinear form to be Hermitian it is necessary and sufficient that $A(\mathbf{x}; \mathbf{x})$ be real for all $\mathbf{x}$.

One example of a Hermitian quadratic form is the form

$$A(\mathbf{x}; \mathbf{x}) = (\mathbf{x}, \mathbf{x}),$$

where $(\mathbf{x}, \mathbf{x})$ denotes the inner product of $\mathbf{x}$ with itself. In fact, axioms 1 through 3 for the inner product in a complex Euclidean space say in effect that $(\mathbf{x}, \mathbf{y})$ is a Hermitian bilinear form so that $(\mathbf{x}, \mathbf{x})$ is a Hermitian quadratic form.

If, as in § 4, we call a quadratic form $A(\mathbf{x}; \mathbf{x})$ positive definite when

$$A(\mathbf{x}; \mathbf{x}) > 0 \qquad \text{for } \mathbf{x} \neq \mathbf{0},$$

then a complex Euclidean space can be defined as a complex vector space with a positive definite Hermitian quadratic form.

If $\mathscr{A}$ is the matrix of a bilinear form $A(\mathbf{x}; \mathbf{y})$ relative to the basis $\mathbf{e}_1, \mathbf{e}_2, \cdots, \mathbf{e}_n$ and $\mathscr{B}$ the matrix of $A(\mathbf{x}; \mathbf{y})$ relative to the basis $\mathbf{f}_1, \mathbf{f}_2, \cdots, \mathbf{f}_n$ and if $\mathbf{f}_j = \sum_{i=1}^{n} c_{ij}\mathbf{e}_i$ $(j = 1, \cdots, n)$, then

$$\mathscr{B} = \mathscr{C}^* \mathscr{A} \mathscr{C}.$$

Here $\mathscr{C} = ||c_{ij}||$ and $\mathscr{C}^* = ||c^*_{ij}||$ is the conjugate transpose of $\mathscr{C}$, i.e., $c^*_{ij} = \bar{c}_{ji}$.

The proof is the same as the proof of the analogous fact in a real space.

5. *Reduction of a quadratic form to a sum of squares*

THEOREM 1. *Let $A(\mathbf{x}; \mathbf{x})$ be a Hermitian quadratic form in a complex vector space* $\mathbf{R}$. *Then there is a basis $\mathbf{e}_1, \mathbf{e}_2, \cdots, \mathbf{e}_n$ of* $\mathbf{R}$ *relative to which the form in question is given by*

$$A(\mathbf{x}; \mathbf{x}) = \lambda_1 \xi_1 \bar{\xi}_1 + \lambda_2 \xi_2 \bar{\xi}_2 + \cdots + \lambda_n \xi_n \bar{\xi}_n,$$

where all the λ's are real.

One can prove the above by imitating the proof in § 5 of the analogous theorem in a real space. We choose to give a version of the proof which emphasizes the geometry of the situation. The idea is to select in succession the vectors of the desired basis.

We choose $\mathbf{e}_1$ so that $A(\mathbf{e}_1; \mathbf{e}_1) \neq 0$. This can be done for otherwise $A(\mathbf{x}; \mathbf{x}) = 0$ for all $\mathbf{x}$ and, in view of formula (1), $A(\mathbf{x}; \mathbf{y}) \equiv 0$. Now we select a vector $\mathbf{e}_2$ in the (n − 1)-dimensional space $\mathbf{R}^{(1)}$ consisting of all vectors $\mathbf{x}$ for which $A(\mathbf{e}_1; \mathbf{x}) = 0$ so that $A(\mathbf{e}_2, \mathbf{e}_2) \neq 0$, etc. This process is continued until we reach the space $\mathbf{R}^{(r)}$ in which $A(\mathbf{x}; \mathbf{y}) \equiv 0$ ($\mathbf{R}^{(r)}$ may consist of the zero vector only). If $\mathbf{R}^{(r)} \neq \mathbf{0}$, then we choose in it some basis $\mathbf{e}_{r+1}$, $\mathbf{e}_{r+2}, \cdots, \mathbf{e}_n$. These vectors and the vectors $\mathbf{e}_1, \mathbf{e}_2, \cdots, \mathbf{e}_r$ form a basis of $\mathbf{R}$.

Our construction implies

$$A(\mathbf{e}_i; \mathbf{e}_k) = 0 \qquad \text{for } i < k.$$

On the other hand, the Hermitian nature of the form $A(\mathbf{x}; \mathbf{y})$ implies

$$A(\mathbf{e}_i; \mathbf{e}_k) = 0 \qquad \text{for } i > k.$$

It follows that if

$$\mathbf{x} = \xi_1 \mathbf{e}_1 + \xi_2 \mathbf{e}_2 + \cdots + \xi_n \mathbf{e}_n$$

is an arbitrary vector, then

$$A(\mathbf{x}; \mathbf{x}) = \xi_1 \bar{\xi}_1 A(\mathbf{e}_1; \mathbf{e}_1) + \xi_2 \bar{\xi}_2 A(\mathbf{e}_2; \mathbf{e}_2) + \cdots + \xi_n \bar{\xi}_n A(\mathbf{e}_n; \mathbf{e}_n),$$

where the numbers $A(\mathbf{e}_i; \mathbf{e}_i)$ are real in view of the Hermitian nature of the quadratic form. If we denote $A(\mathbf{e}_i; \mathbf{e}_i)$ by λ_i, then

$$A(\mathbf{x}; \mathbf{x}) = \lambda_1 \xi_1 \bar{\xi}_1 + \lambda_2 \xi_2 \bar{\xi}_2 + \cdots + \lambda_n \xi_n \bar{\xi}_n = \lambda_1 |\xi_1|^2 + \lambda_2 |\xi_2|^2 + \cdots + \lambda_n |\xi_n|^2.$$

6. *Reduction of a Hermitian quadratic form to a sum of squares by means of a triangular transformation.* Let $A(\mathbf{x}; \mathbf{x})$ be a Hermitian quadratic form in a complex vector space and $\mathbf{e}_1, \mathbf{e}_2, \cdots, \mathbf{e}_n$ a basis. We assume that the determinants

$$\Delta_1 = a_{11}, \quad \Delta_2 = \begin{vmatrix} a_{11} & a_{12} \\ a_{21} & a_{22} \end{vmatrix}, \quad \cdots, \quad \Delta_n = \begin{vmatrix} a_{11} & a_{12} & \cdots & a_{1n} \\ a_{21} & a_{22} & \cdots & a_{2n} \\ \cdots & \cdots & \cdots & \cdots \\ a_{n1} & a_{n2} & \cdots & a_{nn} \end{vmatrix},$$

where $a_{ik} = A(\mathbf{e}_i; \mathbf{e}_k)$, are all different from zero. Then just as in § 6, we can write down formulas for finding a basis relative to which the quadratic form is represented by a sum of squares. These formulas are identical with (3) and (6) of § 6. Relative to such a basis the quadratic form is given by

$$A(\mathbf{x}; \mathbf{x}) = \frac{\Delta_0}{\Delta_1} |\xi_1|^2 + \frac{\Delta_1}{\Delta_2} |\xi_2|^2 + \cdots + \frac{\Delta_{n-1}}{\Delta_n} |\xi_n|^2,$$

where $\Delta_0 = 1$. This implies, among others, that the determinants $\Delta_1, \Delta_2, \cdots, \Delta_n$ are real. To see this we recall that if a Hermitian quadratic form is reduced to the canonical form (3), then the coefficients are equal to $A(\mathbf{e}_i; \mathbf{e}_i)$ and are thus real.

Exercise. Prove directly that if the quadratic form $A(\mathbf{x}; \mathbf{x})$ is Hermitian, then the determinants $\Delta_0, \Delta_1, \cdots, \Delta_n$ are real.

Just as in § 6 we find that *for a Hermitian quadratic form to be positive definite it is necessary and sufficient that the determinants $\Delta_1, \Delta_2, \cdots, \Delta_n$ be positive.*

The number of negative multipliers of the squares in the canonical form of a Hermitian quadratic form equals the number of changes of sign in the sequence

$$1, \Delta_1, \Delta_2, \cdots, \Delta_n.$$

7. *The law of inertia*

Theorem 2. *If a Hermitian quadratic form has canonical form relative to two bases, then the number of positive, negative and zero coefficients is the same in both cases.*

The proof of this theorem is the same as the proof of the corresponding theorem in § 7.

The concept of rank of a quadratic form introduced in § 7 for real spaces can be extended without change to complex spaces.

CHAPTER II

Linear Transformations

§ 9. *Linear transformations. Operations on linear transformations*

1. *Fundamental definitions.* In the preceding chapter we studied functions which associate numbers with points in an n-dimensional vector space. In many cases, however, it is necessary to consider functions which associate points of a vector space with points of that same vector space. The simplest functions of this type are linear transformations.

DEFINITION 1. *If with every vector* $\mathbf{x}$ *of a vector space* $\mathbf{R}$ *there is associated a (unique) vector* $\mathbf{y}$ *in* $\mathbf{R}$, *then the mapping* $\mathbf{y} = A(\mathbf{x})$ *is called a transformation of the space* $\mathbf{R}$.

This transformation is said to be linear if the following two conditions hold:

1. $A(\mathbf{x}_1 + \mathbf{x}_2) = A(\mathbf{x}_1) + A(\mathbf{x}_2)$,
2. $A(\lambda\mathbf{x}) = \lambda A(\mathbf{x})$.

Whenever there is no danger of confusion the symbol $A(\mathbf{x})$ is replaced by the symbol $A\mathbf{x}$.

EXAMPLES. *1.* Consider a rotation of three-dimensional Euclidean space $\mathbf{R}$ about an axis through the origin. If $\mathbf{x}$ is any vector in $\mathbf{R}$, then $A\mathbf{x}$ stands for the vector into which $\mathbf{x}$ is taken by this rotation. It is easy to see that conditions 1 and 2 hold for this mapping. Let us check condition 1, say. The left side of 1 is the result of first adding $\mathbf{x}_1$ and $\mathbf{x}_2$ and then rotating the sum. The right side of 1 is the result of first rotating $\mathbf{x}_1$ and $\mathbf{x}_2$ and then adding the results. Clearly, both procedures yield the same vector.

2. Let $\mathbf{R}'$ be a plane in the space $\mathbf{R}$ (of Example *1*) passing through the origin. We associate with $\mathbf{x}$ in $\mathbf{R}$ its projection $\mathbf{x}' = A\mathbf{x}$ on the plane $\mathbf{R}'$. It is again easy to see that conditions 1 and 2 hold.

3. Consider the vector space of n-tuples of real numbers. Let $\|a_{ik}\|$ be a (square) matrix. With the vector

$$\mathbf{x} = (\xi_1, \xi_2, \cdots, \xi_n)$$

we associate the vector

$$\mathbf{y} = \mathrm{A}\mathbf{x} = (\eta_1, \eta_2, \cdots, \eta_n),$$

where

$$\eta_i = \sum_{k=1}^{n} a_{ik}\xi_k.$$

This mapping is another instance of a linear transformation.

4. Consider the n-dimensional vector space of polynomials of degree $\leqq n - 1$.

If we put

$$\mathrm{A}P(t) = P'(t),$$

where $P'(t)$ is the derivative of $P(t)$, then A is a linear transformation. Indeed

1. $[P_1(t) + P_2(t)]' = P'_1(t) + P'_2(t)$,
2. $[\lambda P(t)]' = \lambda P'(t)$.

5. Consider the space of continuous functions $f(t)$ defined on the interval $[0, 1]$. If we put

$$\mathrm{A}f(t) = \int_0^t f(\tau)\, d\tau,$$

then $\mathrm{A}f(t)$ is a continuous function and A is linear. Indeed,

$$1.\quad \mathrm{A}(f_1 + f_2) = \int_0^t [f_1(\tau) + f_2(\tau)]\, d\tau = \int_0^t f_1(\tau)\, d\tau + \int_0^t f_2(\tau)\, d\tau = \mathrm{A}f_1 + \mathrm{A}f_2;$$

$$2.\quad \mathrm{A}(\lambda f) = \int_0^t \lambda f(\tau)\, d\tau = \lambda \int_0^t f(\tau)\, d\tau = \lambda \mathrm{A}f.$$

Among linear transformations the following simple transformations play a special role.

The identity mapping E defined by the equation

$$\mathrm{E}\mathbf{x} = \mathbf{x}$$

for all $\mathbf{x}$.

The null transformation O defined by the equation

$$\mathrm{O}\mathbf{x} = \mathbf{0}$$

for all $\mathbf{x}$.

2. *Connection between matrices and linear transformations.* Let $\mathbf{e}_1, \mathbf{e}_2, \cdots, \mathbf{e}_n$ be a basis of an n-dimensional vector space $\mathbf{R}$ and let A denote a linear transformation on $\mathbf{R}$. We shall show that

Given n arbitrary vectors $\mathbf{g}_1, \mathbf{g}_2, \cdots, \mathbf{g}_n$ there exists a unique linear transformation A *such that*

$$\mathrm{A}\mathbf{e}_1 = \mathbf{g}_1, \quad \mathrm{A}\mathbf{e}_2 = \mathbf{g}_2, \quad \cdots, \quad \mathrm{A}\mathbf{e}_n = \mathbf{g}_n.$$

We first prove that the vectors $\mathrm{A}\mathbf{e}_1, \mathrm{A}\mathbf{e}_2, \cdots, \mathrm{A}\mathbf{e}_n$ determine A uniquely. In fact, if

$$\mathbf{x} = \xi_1\mathbf{e}_1 + \xi_2\mathbf{e}_2 + \cdots + \xi_n\mathbf{e}_n \tag{1}$$

is an arbitrary vector in $\mathbf{R}$, then

$$\mathrm{A}\mathbf{x} = \mathrm{A}(\xi_1\mathbf{e}_1 + \xi_2\mathbf{e}_2 + \cdots + \xi_n\mathbf{e}_n) = \xi_1\mathrm{A}\mathbf{e}_1 + \xi_2\mathrm{A}\mathbf{e}_2 + \cdots + \xi_n\mathrm{A}\mathbf{e}_n, \tag{2}$$

so that A is indeed uniquely determined by the $\mathrm{A}\mathbf{e}_i$.

It remains to prove the existence of A with the desired properties. To this end we consider the mapping A which associates with $\mathbf{x} = \xi_1\mathbf{e}_1 + \xi_2\mathbf{e}_2 + \cdots + \xi_n\mathbf{e}_n$ the vector $\mathrm{A}\mathbf{x} = \xi_1\mathbf{g}_1 + \xi_2\mathbf{g}_2 + \cdots + \xi_n\mathbf{g}_n$. This mapping is well defined, since $\mathbf{x}$ has a unique representation relative to the basis $\mathbf{e}_1, \mathbf{e}_2, \cdots, \mathbf{e}_n$. It is easily seen that the mapping A is linear.

Now let the coordinates of $\mathbf{g}_k$ relative to the basis $\mathbf{e}_1, \mathbf{e}_2, \cdots, \mathbf{e}_n$ be $a_{1k}, a_{2k}, \cdots, a_{nk}$, i.e.,

$$\mathbf{g}_k = \mathrm{A}\mathbf{e}_k = \sum_{i=1}^{n} a_{ik}\mathbf{e}_i. \tag{3}$$

The numbers a_{ik} $(i, k = 1, 2, \cdots, n)$ form a matrix

$$\mathscr{A} = ||a_{ik}||$$

which we shall call *the matrix of the linear transformation* A *relative to the basis* $\mathbf{e}_1, \mathbf{e}_2, \cdots, \mathbf{e}_n$.

We have thus shown that relative to a given basis $\mathbf{e}_1, \mathbf{e}_2, \cdots, \mathbf{e}_n$ *every linear transformation* A *determines a unique matrix* $||a_{ik}||$ *and, conversely, every matrix determines a unique linear transformation given by means of the formulas* (3), (1), (2).

Linear transformations can thus be described by means of matrices and matrices are the analytical tools for the study of linear transformations on vector spaces.

EXAMPLES. *1.* Let **R** be the three-dimensional Euclidean space and A the linear transformation which projects every vector on the XY-plane. We choose as basis vectors of **R** unit vectors $\mathbf{e}_1$, $\mathbf{e}_2$, $\mathbf{e}_3$ directed along the coordinate axes. Then

$$\mathrm{A}\mathbf{e}_1 = \mathbf{e}_1, \quad \mathrm{A}\mathbf{e}_2 = \mathbf{e}_2, \quad \mathrm{A}\mathbf{e}_3 = \mathbf{0},$$

i.e., relative to this basis the mapping A is represented by the matrix

$$\begin{bmatrix} 1 & 0 & 0 \\ 0 & 1 & 0 \\ 0 & 0 & 0 \end{bmatrix}.$$

EXERCISE. Find the matrix of the above transformation relative to the basis $\mathbf{e}'_1$, $\mathbf{e}'_2$, $\mathbf{e}'_3$, where

$$\mathbf{e}'_1 = \mathbf{e}_1, \quad \mathbf{e}'_2 = \mathbf{e}_2, \mathbf{e}'_3 = \mathbf{e}_1 + \mathbf{e}_2 + \mathbf{e}_3.$$

2. Let E be the identity mapping and $\mathbf{e}_1, \mathbf{e}_2, \cdots, \mathbf{e}_n$ any basis in **R**. Then

$$\mathrm{A}\mathbf{e}_i = \mathbf{e}_i \ (i = 1, 2, \cdots, n),$$

i.e., the matrix which represents E relative to any basis is

$$\begin{bmatrix} 1 & 0 & \cdots & 0 \\ 0 & 1 & \cdots & 0 \\ \cdots & \cdots & \cdots & \cdots \\ 0 & 0 & \cdots & 1 \end{bmatrix}.$$

It is easy to see that the null transformation is always represented by the matrix all of whose entries are zero.

3. Let **R** be the space of polynomials of degree $\leqq n - 1$. Let A be the differentiation transformation, i.e.,

$$\mathrm{A}P(t) = P'(t).$$

We choose the following basis in **R**:

$$\mathbf{e}_1 = 1, \qquad \mathbf{e}_2 = t, \qquad \mathbf{e}_3 = \frac{t^2}{2!}, \qquad \cdots, \qquad \mathbf{e}_n = \frac{t^{n-1}}{(n-1)!}.$$

Then

$$\mathrm{A}\mathbf{e}_1 = 1' = 0, \qquad \mathrm{A}\mathbf{e}_2 = t' = 1 = \mathbf{e}_1, \qquad \mathrm{A}\mathbf{e}_3 = \left(\frac{t^2}{2}\right)' = t = \mathbf{e}_2,$$

$$\cdots, \qquad \mathrm{A}\mathbf{e}_n = \left(\frac{t^{n-1}}{(n-1)!}\right)' = \frac{t^{n-2}}{(n-2)!} = \mathbf{e}_{n-1}.$$

Hence relative to our basis, A is represented by the matrix

$$\begin{bmatrix} 0 & 1 & 0 & \cdots & 0 \\ 0 & 0 & 1 & \cdots & 0 \\ \cdot & \cdot & \cdot & \cdots & \cdot \\ 0 & 0 & 0 & \cdots & 1 \\ 0 & 0 & 0 & \cdots & 0 \end{bmatrix}.$$

Let A be a linear transformation, $\mathbf{e}_1, \mathbf{e}_2, \cdots, \mathbf{e}_n$ a basis in $\mathbf{R}$ and $\|a_{ik}\|$ the matrix which represents A relative to this basis. Let

$$\mathbf{x} = \xi_1\mathbf{e}_1 + \xi_2\mathbf{e}_2 + \cdots + \xi_n\mathbf{e}_n, \tag{4}$$

$$\mathrm{A}\mathbf{x} = \eta_1\mathbf{e}_1 + \eta_2\mathbf{e}_2 + \cdots + \eta_n\mathbf{e}_n. \tag{4'}$$

We wish to express the coordinates η_i of $\mathrm{A}\mathbf{x}$ by means of the coordinates ξ_i of $\mathbf{x}$. Now

$$\begin{aligned} \mathrm{A}\mathbf{x} &= \mathrm{A}(\xi_1\mathbf{e}_1 + \xi_2\mathbf{e}_2 + \cdots + \xi_n\mathbf{e}_n) \\ &= \xi_1(a_{11}\mathbf{e}_1 + a_{21}\mathbf{e}_2 + \cdots + a_{n1}\mathbf{e}_n) \\ &+ \xi_2(a_{12}\mathbf{e}_1 + a_{22}\mathbf{e}_2 + \cdots + a_{n2}\mathbf{e}_2) \\ &+ \cdots\cdots\cdots\cdots\cdots \\ &+ \xi_n(a_{1n}\mathbf{e}_1 + a_{2n}\mathbf{e}_2 + \cdots + a_{nn}\mathbf{e}_n) \\ &= (a_{11}\xi_1 + a_{12}\xi_2 + \cdots + a_{1n}\xi_n)\mathbf{e}_1 \\ &+ (a_{21}\xi_1 + a_{22}\xi_2 + \cdots + a_{2n}\xi_n)\mathbf{e}_2 \\ &+ \cdots\cdots\cdots\cdots\cdots \\ &+ (a_{n1}\xi_1 + a_{n2}\xi_2 + \cdots + a_{nn}\xi_n)\mathbf{e}_n. \end{aligned}$$

Hence, in view of (4'),

$$\begin{aligned} \eta_1 &= a_{11}\xi_1 + a_{12}\xi_2 + \cdots + a_{1n}\xi_n, \\ \eta_2 &= a_{12}\xi_1 + a_{22}\xi_2 + \cdots + a_{2n}\xi_n, \\ &\cdots\cdots\cdots\cdots\cdots \\ \eta_n &= a_{n1}\xi_1 + a_{n2}\xi_2 + \cdots + a_{nn}\xi_n, \end{aligned}$$

or, briefly,

$$\eta_i = \sum_{k=1}^{n} a_{ik}\xi_k \tag{5}$$

Thus, *if* $||a_{ik}||$ *represents a linear transformation* A *relative to some basis* $\mathbf{e}_1, \mathbf{e}_2, \cdots, \mathbf{e}_n$, *then transformation of the basis vectors involves the columns of* $||a_{ik}||$ [*formula* (3)] *and transformation of the coordinates of an arbitrary vector* $\mathbf{x}$ *involves the rows of* $||a_{ik}||$ [*formula* (5)].

3. *Addition and multiplication of linear transformations.* We shall now define addition and multiplication for linear transformations.

DEFINITION 2. *By the product of two linear transformations* A *and* B *we mean the transformation* C *defined by the equation* $\mathrm{C}\mathbf{x} = \mathrm{A}(\mathrm{B}\mathbf{x})$ *for all* $\mathbf{x}$.

If C is the product of A and B, we write C = AB.

The product of linear transformations is itself linear, i.e., it satisfies conditions 1 and 2 of Definition 1. Indeed,

$$\begin{aligned} \mathrm{C}(\mathbf{x}_1 + \mathbf{x}_2) &= \mathrm{A}[\mathrm{B}(\mathbf{x}_1 + \mathbf{x}_2)] = \mathrm{A}(\mathrm{B}\mathbf{x}_1 + \mathrm{B}\mathbf{x}_2) \\ &= \mathrm{AB}\mathbf{x}_1 + \mathrm{AB}\mathbf{x}_2 = \mathrm{C}\mathbf{x}_1 + \mathrm{C}\mathbf{x}_2. \end{aligned}$$

The first equality follows from the definition of multiplication of transformations, the second from property 1 for B, the third from property 1 for A and the fourth from the definition of multiplication of transformations. That $\mathrm{C}(\lambda\mathbf{x}) = \lambda\mathrm{C}\mathbf{x}$ is proved just as easily.

If E is the identity transformation and A is an arbitrary transformation, then it is easy to verify the relations

$$\mathrm{AE} = \mathrm{EA} = \mathrm{A}.$$

Next we define powers of a transformation A:

$$\mathrm{A}^2 = \mathrm{A} \cdot \mathrm{A}, \qquad \mathrm{A}^3 = \mathrm{A}^2 \cdot \mathrm{A}, \qquad \cdots \text{ etc.},$$

and, by analogy with numbers, we define $\mathrm{A}^0 = \mathrm{E}$. Clearly,

$$\mathrm{A}^{m+n} = \mathrm{A}^m \cdot \mathrm{A}^n.$$

EXAMPLE. Let $\mathbf{R}$ be the space of polynomials of degree $\leqq n - 1$. Let D be the differentiation operator,

$$\mathrm{D}P(t) = P'(t).$$

Then $\mathrm{D}^2P(t) = \mathrm{D}(\mathrm{D}P(t)) = (P'(t))' = P''(t)$. Likewise, $\mathrm{D}^3P(t) = P'''(t)$. Clearly, in this case $\mathrm{D}^n = \mathrm{O}$.

EXERCISE. Select in $\mathbf{R}$ of the above example a basis as in Example *3* of para. 3 of this section and find the matrices of D, D^2, D^3, $\cdots$ relative to this basis.

We know that given a basis $\mathbf{e}_1, \mathbf{e}_2, \cdots, \mathbf{e}_n$ every linear transformation determines a matrix. If the transformation A determines the matrix $||a_{ik}||$ and B the matrix $||b_{ik}||$, what is the matrix $||c_{ik}||$ determined by the product C of A and B. To answer this question we note that by definition of $||c_{ik}||$

$$\mathrm{C}\mathbf{e}_k = \sum_i c_{ik}\mathbf{e}_i. \tag{6}$$

Further

$$\mathrm{AB}\mathbf{e}_k = \mathrm{A}(\sum_{j=1}^{n} b_{jk}\mathbf{e}_j) = \sum_j b_{jk}\mathrm{A}\mathbf{e}_j = \sum_{i,j} b_{jk}a_{ij}\mathbf{e}_i. \tag{7}$$

Comparison of (7) and (6) yields

$$c_{ik} = \sum_j a_{ij}b_{jk}. \tag{8}$$

We see that the element c_{ik} of the matrix $\mathscr{C}$ is the sum of the products of the elements of the ith row of the matrix $\mathscr{A}$ and the corresponding elements of the kth column of the matrix $\mathscr{B}$. The matrix $\mathscr{C}$ with entries defined by (8) is called the product of the matrices $\mathscr{A}$ and $\mathscr{B}$ in this order. Thus, if the (linear) transformation A is represented by the matrix $||a_{ik}||$ and the (linear) transformation B by the matrix $||b_{ik}||$, then their product is represented by the matrix $||c_{ik}||$ which is the product of the matrices $||a_{ik}||$ and $||b_{ik}||$.

DEFINITION 3. *By the sum of two linear transformations* A *and* B *we mean the transformation* C *defined by the equation* $\mathrm{C}\mathbf{x} = \mathrm{A}\mathbf{x} + \mathrm{B}\mathbf{x}$ *for all* $\mathbf{x}$.

If C is the sum of A and B we write $\mathrm{C} = \mathrm{A} + \mathrm{B}$. It is easy to see that C is linear.

Let C be the sum of the transformations A and B. If $||a_{ik}||$ and $||b_{ik}||$ represent A and B respectively (relative to some basis $\mathbf{e}_1, \mathbf{e}_2, \cdots, \mathbf{e}_n$) and $||c_{ik}||$ represents the sum C of A and B (relative to the same basis), then, on the one hand,

$$\mathrm{A}\mathbf{e}_k = \sum_i a_{ik}\mathbf{e}_i, \qquad \mathrm{B}\mathbf{e}_k = \sum_i b_{ik}\mathbf{e}_i, \qquad \mathrm{C}\mathbf{e}_k = \sum_i c_{ik}\mathbf{e}_i,$$

and, on the other hand,

$$\mathrm{C}\mathbf{e}_k = \mathrm{A}\mathbf{e}_k + \mathrm{B}\mathbf{e}_k = \sum_i (a_{ik} + b_{ik})\mathbf{e}_i,$$

so that

$$c_{ik} = a_{ik} + b_{ik}.$$

The matrix $||a_{ik} + b_{ik}||$ is called the sum of the matrices $||a_{ik}||$ and $||b_{ik}||$. Thus *the matrix of the sum of two linear transformations is the sum of the matrices associated with the summands.*

Addition and multiplication of linear transformations have some of the properties usually associated with these operations. Thus

1. $\mathrm{A} + \mathrm{B} = \mathrm{B} + \mathrm{A}$;
2. $(\mathrm{A} + \mathrm{B}) + \mathrm{C} = \mathrm{A} + (\mathrm{B} + \mathrm{C})$;
3. $\mathrm{A}(\mathrm{BC}) = (\mathrm{AB})\mathrm{C}$;
4. $\begin{cases} (\mathrm{A} + \mathrm{B})\mathrm{C} = \mathrm{AC} + \mathrm{BC}, \\ \mathrm{C}(\mathrm{A} + \mathrm{B}) = \mathrm{CA} + \mathrm{CB}. \end{cases}$

We could easily prove these equalities directly but this is unnecessary. We recall that we have established the existence of a one-to-one correspondence between linear transformations and matrices which preserves sums and products. Since properties 1 through 4 are proved for matrices in a course in algebra, the isomorphism between matrices and linear transformations just mentioned allows us to claim the validity of 1 through 4 for linear transformations.

We now define the product of a number λ and a linear transformation A. Thus by $\lambda\mathrm{A}$ we mean the transformation which associates with every vector $\mathbf{x}$ the vector $\lambda(\mathrm{A}\mathbf{x})$. It is clear that if A is represented by the matrix $||a_{ik}||$, then $\lambda\mathrm{A}$ is represented by the matrix $||\lambda a_{ik}||$.

If $P(t) = a_0 t^m + a_1 t^{m-1} + \cdots + a_m$ is an arbitrary polynomial and A is a transformation, we define the symbol $P(\mathrm{A})$ by the equation

$$P(\mathrm{A}) = a_0 \mathrm{A}^m + a_1 \mathrm{A}^{m-1} + \cdots + a_m \mathrm{E}.$$

EXAMPLE. Consider the space $\mathbf{R}$ of functions defined and infinitely differentiable on an interval (a, b). Let D be the linear mapping defined on $\mathbf{R}$ by the equation

$$\mathrm{D}f(t) = f'(t).$$

If $P(t)$ is the polynomial $P(t) = a_0 t^m + a_1 t^{m-1} + \cdots + a_m$, then $P(\mathrm{D})$ is the linear mapping which takes $f(t)$ in $\mathbf{R}$ into

$$P(\mathrm{D})f(t) = a_0 f^{(m)}(t) + a_1 f^{(m-1)}(t) + \cdots + a_m f(t).$$

Analogously, with $P(t)$ as above and $\mathscr{A}$ a matrix we define $P(\mathscr{A})$, a polynomial in a matrix, by means of the equation

$$P(\mathscr{A}) = a_0 \mathscr{A}^m + a_1 \mathscr{A}^{m-1} + \cdots + a_0 \mathscr{E}.$$

EXAMPLE. Let $\mathscr{A}$ be a diagonal matrix, i.e., a matrix of the form

$$\mathscr{A} = \begin{bmatrix} \lambda_1 & 0 & 0 & \cdots & 0 \\ 0 & \lambda_2 & 0 & \cdots & 0 \\ \cdots & \cdots & \cdots & \cdots & \cdots \\ 0 & 0 & 0 & \cdots & \lambda_n \end{bmatrix}.$$

We wish to find $P(\mathscr{A})$. Since

$$\mathscr{A}^2 = \begin{bmatrix} \lambda_1^2 & 0 & \cdots & 0 \\ 0 & \lambda_2^2 & \cdots & 0 \\ \cdots & \cdots & \cdots & \cdots \\ 0 & 0 & \cdots & \lambda_n^2 \end{bmatrix}, \quad \cdots, \quad \mathscr{A}^m = \begin{bmatrix} \lambda_1^m & 0 & \cdots & 0 \\ 0 & \lambda_2^m & \cdots & 0 \\ \cdots & \cdots & \cdots & \cdots \\ 0 & 0 & \cdots & \lambda_n^m \end{bmatrix},$$

it follows that

$$P(\mathscr{A}) = \begin{bmatrix} P(\lambda_1) & 0 & \cdots & 0 \\ 0 & P(\lambda_2) & \cdots & 0 \\ \cdots & \cdots & \cdots & \cdots \\ 0 & 0 & \cdots & P(\lambda_n) \end{bmatrix}.$$

EXERCISE. Find $P(\mathscr{A})$ for

$$\mathscr{A} = \begin{bmatrix} 0 & 1 & 0 & 0 & \cdots & 0 \\ 0 & 0 & 1 & 0 & \cdots & 0 \\ 0 & 0 & 0 & 1 & \cdots & 0 \\ \cdots & \cdots & \cdots & \cdots & \cdots & \cdots \\ 0 & 0 & 0 & 0 & \cdots & 1 \\ 0 & 0 & 0 & 0 & \cdots & 0 \end{bmatrix}.$$

It is possible to give reasonable definitions not only for a polynomial in a matrix $\mathscr{A}$ but also for any function of a matrix $\mathscr{A}$ such as $\exp \mathscr{A}$, $\sin \mathscr{A}$, etc.

As was already mentioned in § 1, Example 5, all matrices of order n with the usual definitions of addition and multiplication by a scalar form a vector space of dimension n^2. Hence any $n^2 + 1$ matrices are linearly dependent. Now consider the following set of powers of some matrix $\mathscr{A}$:

$$\mathscr{E}, \mathscr{A}, \mathscr{A}^2, \cdots, \mathscr{A}^{n^2}.$$

Since the number of matrices is $n^2 + 1$, they must be linearly dependent, that is, there exist numbers $a_0, a_1, a_2, \cdots, a_{n^2}$ (not all zero) such that

$$a_0\mathscr{E} + a_1\mathscr{A} + a_2\mathscr{A}^2 + \cdots + a_{n^2}\mathscr{A}^{n^2} = \mathscr{O}.$$

It follows that for every matrix of order n there exists a polynomial P of degree at most n^2 such that $P(\mathscr{A}) = \mathscr{O}$. This simple proof of the existence of a polynomial $P(t)$ for which $P(\mathscr{A}) = \mathscr{O}$ is deficient in two respects, namely, it does not tell us how to construct $P(t)$ and it suggests that the degree of $P(t)$ may be as high as n^2. In the sequel we shall prove that for every matrix $\mathscr{A}$ there exists a polynomial $P(t)$ of degree n derivable in a simple manner from $\mathscr{A}$ and having the property $P(\mathscr{A}) = \mathscr{O}$.

4. *Inverse transformation*

DEFINITION 4. *The transformation* B *is said to be the inverse of* A *if* AB = BA = E, *where* E *is the identity mapping.*

The definition implies that $\mathrm{B}(\mathrm{A}\mathbf{x}) = \mathbf{x}$ for all $\mathbf{x}$, i.e., if A takes $\mathbf{x}$ into $\mathrm{A}\mathbf{x}$, then the inverse B of A takes $\mathrm{A}\mathbf{x}$ into $\mathbf{x}$. The inverse of A is usually denoted by A^{-1}.

Not every transformation possesses an inverse. Thus it is clear that the projection of vectors in three-dimensional Euclidean space on the XY-plane has no inverse.

There is a close connection between the inverse of a transformation and the inverse of a matrix. As is well-known for every matrix $\mathscr{A}$ with non-zero determinant there exists a matrix $\mathscr{A}^{-1}$ such that

$$\mathscr{A}\mathscr{A}^{-1} = \mathscr{A}^{-1}\mathscr{A} = \mathscr{E}. \tag{9}$$

$\mathscr{A}^{-1}$ is called the inverse of $\mathscr{A}$. To find $\mathscr{A}$ we must solve a system of linear equations equivalent to the matrix equation (9). The elements of the kth column of $\mathscr{A}^{-1}$ turn out to be the cofactors of the elements of the kth row of $\mathscr{A}$ divided by the determinant of $\mathscr{A}$. It is easy to see that $\mathscr{A}^{-1}$ as just defined satisfies equation (9).

We know that choice of a basis determines a one-to-one correspondence between linear transformations and matrices which preserves products. It follows that a *linear transformation* A *has an inverse if and only if its matrix relative to any basis has a non-zero determinant, i.e., the matrix has rank n.* A transformation which has an inverse is sometimes called non-singular.

If A is a singular transformation, then its matrix has rank $< n$.

We shall prove that the rank of the matrix of a linear transformation is independent of the choice of basis.

THEOREM. *Let* A *be a linear transformation on a space* $\mathbf{R}$. *The set of vectors* $A\mathbf{x}$ ($\mathbf{x}$ *varies on* $\mathbf{R}$) *forms a subspace* $\mathbf{R}'$ *of* $\mathbf{R}$. *The dimension of* $\mathbf{R}'$ *equals the rank of the matrix of* A *relative to any basis* $\mathbf{e}_1, \mathbf{e}_2, \cdots, \mathbf{e}_n$.

Proof: Let $\mathbf{y}_1 \in \mathbf{R}'$ and $\mathbf{y}_2 \in \mathbf{R}'$, i.e., $\mathbf{y}_1 = A\mathbf{x}_1$ and $\mathbf{y}_2 = A\mathbf{x}_2$. Then

$$\mathbf{y}_1 + \mathbf{y}_2 = A\mathbf{x}_1 + A\mathbf{x}_2 = A(\mathbf{x}_1 + \mathbf{x}_2),$$

i.e., $\mathbf{y}_1 + \mathbf{y}_2 \in \mathbf{R}'$. Likewise, if $\mathbf{y} = A\mathbf{x}$, then

$$\lambda\mathbf{y} = \lambda A\mathbf{x} = A(\lambda\mathbf{x}),$$

i.e., $\lambda\mathbf{y} \in \mathbf{R}'$. Hence $\mathbf{R}'$ is indeed a subspace of $\mathbf{R}$.

Now any vector $\mathbf{x}$ is a linear combination of the vectors $\mathbf{e}_1, \mathbf{e}_2, \cdots, \mathbf{e}_n$. Hence every vector $A\mathbf{x}$, i.e., every vector in $\mathbf{R}'$, is a linear combination of the vectors $A\mathbf{e}_1, A\mathbf{e}_2, \cdots, A\mathbf{e}_n$. If the maximal number of linearly independent vectors among the $A\mathbf{e}_i$ is k, then the other $A\mathbf{e}_i$ are linear combinations of the k vectors of such a maximal set. Since every vector in $\mathbf{R}'$ is a linear combination of the vectors $A\mathbf{e}_1, A\mathbf{e}_2, \cdots, A\mathbf{e}_n$, it is also a linear combination of the k vectors of a maximal set. Hence the dimension of $\mathbf{R}'$ is k. Let $||a_{ij}||$ represent A relative to the basis $\mathbf{e}_1, \mathbf{e}_2, \cdots, \mathbf{e}_n$. To say that the maximal number of linearly independent $A\mathbf{e}_i$ is k is to say that the maximal number of linearly independent columns of the matrix $||a_{ij}||$ is k, i.e., the dimension of $\mathbf{R}'$ is the same as the rank of the matrix $||a_{ij}||$.

5. *Connection between the matrices of a linear transformation relative to different bases.* The matrices which represent a linear transformation in different bases are usually different. We now show how the matrix of a linear transformation changes under a change of basis.

Let $\mathbf{e}_1, \mathbf{e}_2, \cdots, \mathbf{e}_n$ and $\mathbf{f}_1, \mathbf{f}_2, \cdots, \mathbf{f}_n$ be two bases in $\mathbf{R}$. Let $\mathscr{C}$ be the matrix connecting the two bases. More specifically, let

$$\begin{aligned} \mathbf{f}_1 &= c_{11}\mathbf{e}_1 + c_{21}\mathbf{e}_2 + \cdots + c_{n1}\mathbf{e}_n, \\ \mathbf{f}_2 &= c_{12}\mathbf{e}_1 + c_{22}\mathbf{e}_2 + \cdots + c_{n2}\mathbf{e}_n, \\ &\cdots\cdots\cdots\cdots\cdots\cdots\cdots\cdots \\ \mathbf{f}_n &= c_{1n}\mathbf{e}_1 + c_{2n}\mathbf{e}_2 + \cdots + c_{nn}\mathbf{e}_n. \end{aligned} \tag{10}$$

If C is the linear transformation defined by the equations

$$C\mathbf{e}_i = \mathbf{f}_i \quad (i = 1, 2, \cdots, n),$$

then the matrix of C relative to the basis $\mathbf{e}_1, \mathbf{e}_2, \cdots, \mathbf{e}_n$ is $\mathscr{C}$ (cf. formulas (2) and (3) of para. 3).

Let $\mathscr{A} = ||a_{ik}||$ be the matrix of A relative to $\mathbf{e}_1, \mathbf{e}_2, \cdots, \mathbf{e}_n$ and $\mathscr{B} = ||b_{ik}||$ its matrix relative to $\mathbf{f}_1, \mathbf{f}_2, \cdots, \mathbf{f}_n$. In other words,

$$\text{(10')} \qquad \mathrm{A}\mathbf{e}_k = \sum_{i=1}^{n} a_{ik}\mathbf{e}_i,$$

$$\text{(10'')} \qquad \mathrm{A}\mathbf{f}_k = \sum_{i=1}^{n} b_{ik}\mathbf{f}_i.$$

We wish to express the matrix $\mathscr{B}$ in terms of the matrices $\mathscr{A}$ and $\mathscr{C}$. To this end we rewrite (10″) as

$$\mathrm{AC}\mathbf{e}_k = \sum_{i=1}^{n} b_{ik}\mathrm{C}\mathbf{e}_i.$$

Premultiplying both sides of this equation by C^{-1} (which exists in view of the linear independence of the $\mathbf{f}_i$) we get

$$\mathrm{C}^{-1}\mathrm{AC}\mathbf{e}_k = \sum_{i=1}^{n} b_{ik}\mathbf{e}_i.$$

It follows that the matrix $||b_{ik}||$ represents $\mathrm{C}^{-1}\mathrm{AC}$ relative to the basis $\mathbf{e}_1, \mathbf{e}_2, \cdots, \mathbf{e}_n$. However, relative to a given basis matrix $(\mathrm{C}^{-1}\mathrm{AC})$ = matrix (C^{-1}) · matrix (A) · matrix (C), so that

$$\text{(11)} \qquad \mathscr{B} = \mathscr{C}^{-1}\mathscr{A}\mathscr{C}.$$

To sum up: *Formula* (11) *gives the connection between the matrix* $\mathscr{B}$ *of a transformation* A *relative to a basis* $\mathbf{f}_1, \mathbf{f}_2, \cdots, \mathbf{f}_n$ *and the matrix* $\mathscr{A}$ *which represents* A *relative to the basis* $\mathbf{e}_1, \mathbf{e}_2, \cdots, \mathbf{e}_n$. *The matrix* $\mathscr{C}$ *in* (11) *is the matrix of transition from the basis* $\mathbf{e}_1, \mathbf{e}_2, \cdots, \mathbf{e}_n$ *to the basis* $\mathbf{f}_1, \mathbf{f}_2, \cdots, \mathbf{f}_n$ (*formula* (10)).

§ 10. *Invariant subspaces. Eigenvalues and eigenvectors of a linear transformation*

1. *Invariant subspaces.* In the case of a scalar valued function defined on a vector space $\mathbf{R}$ but of interest only on a subspace $\mathbf{R}_1$ of $\mathbf{R}$ we may, of course, consider the function on the subspace $\mathbf{R}_1$ only.

Not so in the case of linear transformations. Here points in $\mathbf{R}_1$ may be mapped on points not in $\mathbf{R}_1$ and in that case it is not possible to restrict ourselves to $\mathbf{R}_1$ alone.

DEFINITION 1. *Let* A *be a linear transformation on a space* **R**. *A subspace* $\mathbf{R}_1$ *of* **R** *is called invariant under* A *if* $\mathbf{x} \in \mathbf{R}_1$ *implies* $A\mathbf{x} \in \mathbf{R}_1$.

If a subspace $\mathbf{R}_1$ is invariant under a linear transformation A we may, of course, consider A on $\mathbf{R}_1$ only.

Trivial examples of invariant subspaces are the subspace consisting of the zero element only and the whole space.

EXAMPLES. *1.* Let **R** be three-dimensional Euclidean space and A a rotation about an axis through the origin. The invariant subspaces are: the axis of rotation (a one-dimensional invariant subspace) and the plane through the origin and perpendicular to the axis of rotation (a two-dimensional invariant subspace).

2. Let **R** be a plane. Let A be a stretching by a factor λ_1 along the x-axis and by a factor λ_2 along the y-axis, i.e., A is the mapping which takes the vector $\mathbf{z} = \xi_1 \mathbf{e}_1 + \xi_2 \mathbf{e}_2$ into the vector $A\mathbf{z} = \lambda_1 \xi_1 \mathbf{e}_1 + \lambda_2 \xi_2 \mathbf{e}_2$ (here $\mathbf{e}_1$ and $\mathbf{e}_2$ are unit vectors along the coordinate axes). In this case the coordinate axes are one-dimensional invariant subspaces. If $\lambda_1 = \lambda_2 = \lambda$, then A is a similarity transformation with coefficient λ. In this case every line through the origin is an invariant subspace.

EXERCISE. Show that if $\lambda_1 \neq \lambda_2$, then the coordinate axes are the only invariant one-dimensional subspaces.

3. Let **R** be the space of polynomials of degree $\leqq n - 1$ and A the differentiation operator on **R**, i.e.,

$$AP(t) = P'(t).$$

The set of polynomials of degree $\leqq k \leqq n - 1$ is an invariant subspace.

EXERCISE. Show that **R** in Example *3* contains no other subspaces invariant under A.

4. Let **R** be any n-dimensional vector space. Let A be a linear transformation on **R** whose matrix relative to some basis $\mathbf{e}_1, \mathbf{e}_2, \cdots, \mathbf{e}_n$ is of the form

$$\begin{bmatrix} a_{11} & \cdots & a_{1k} & a_{1k+1} & \cdots & a_{1n} \\ \cdots & \cdots & \cdots & \cdots & \cdots & \cdots \\ a_{k1} & \cdots & a_{kk} & a_{kk+1} & \cdots & a_{kn} \\ 0 & \cdots & 0 & a_{k+1\,k+1} & \cdots & a_{k+1n} \\ \cdots & \cdots & \cdots & \cdots & \cdots & \cdots \\ 0 & \cdots & 0 & a_{nk+1} & \cdots & a_{nn} \end{bmatrix}$$

In this case the subspace generated by the vectors $\mathbf{e}_1, \mathbf{e}_2, \cdots, \mathbf{e}_k$ is invariant under A. The proof is left to the reader. If

$$a_{ik+1} = \cdots = a_{in} = 0 \qquad (1 \leqq i \leqq k),$$

then the subspace generated by $\mathbf{e}_{k+1}, \mathbf{e}_{k+2}, \cdots, \mathbf{e}_n$ would also be invariant under A.

2. *Eigenvectors and eigenvalues.* In the sequel one-dimensional invariant subspaces will play a special role.

Let $\mathbf{R}_1$ be a one-dimensional subspace generated by some vector $\mathbf{x} \neq \mathbf{0}$. Then $\mathbf{R}_1$ consists of all vectors of the form $\alpha\mathbf{x}$. It is clear that for $\mathbf{R}_1$ to be invariant it is necessary and sufficient that the vector $A\mathbf{x}$ be in $\mathbf{R}_1$, i.e., that

$$A\mathbf{x} = \lambda\mathbf{x}.$$

DEFINITION 2. *A vector* $\mathbf{x} \neq \mathbf{0}$ *satisfying the relation* $A\mathbf{x} = \lambda\mathbf{x}$ *is called an eigenvector of* A. *The number* λ *is called an eigenvalue of* A.

Thus if $\mathbf{x}$ is an eigenvector, then the vectors $\alpha\mathbf{x}$ form a one-dimensional invariant subspace.

Conversely, all non-zero vectors of a one-dimensional invariant subspace are eigenvectors.

THEOREM 1. *If* A *is a linear transformation on a complex*[1] *space* $\mathbf{R}$, *then* A *has at least one eigenvector.*

Proof: Let $\mathbf{e}_1, \mathbf{e}_2, \cdots, \mathbf{e}_n$ be a basis in $\mathbf{R}$. Relative to this basis A is represented by some matrix $||a_{ik}||$. Let

$$\mathbf{x} = \xi_1\mathbf{e}_1 + \xi_2\mathbf{e}_2 + \cdots + \xi_n\mathbf{e}_n$$

be any vector in $\mathbf{R}$. Then the coordinates $\eta_1, \eta_2, \cdots, \eta_n$ of the vector $A\mathbf{x}$ are given by

[1] The proof holds for a vector space over any algebraically closed field since it makes use only of the fact that equation (2) has a solution.

$$\begin{aligned}\eta_1 &= a_{11}\xi_1 + a_{12}\xi_2 + \cdots + a_{1n}\xi_n, \\ \eta_2 &= a_{21}\xi_1 + a_{22}\xi_2 + \cdots + a_{2n}\xi_n, \\ &\cdots\cdots\cdots\cdots\cdots\cdots \\ \eta_n &= a_{n1}\xi_1 + a_{n2}\xi_2 + \cdots + a_{nn}\xi_n\end{aligned}$$

(Cf. para. 3 of § 9).
The equation

$$\mathbf{Ax} = \lambda\mathbf{x},$$

which expresses the condition for $\mathbf{x}$ to be an eigenvector, is equivalent to the system of equations:

$$\begin{aligned}a_{11}\xi_1 + a_{12}\xi_2 + \cdots + a_{1n}\xi_n &= \lambda\xi_1, \\ a_{21}\xi_1 + a_{22}\xi_2 + \cdots + a_{2n}\xi_n &= \lambda\xi_2, \\ \cdots\cdots\cdots\cdots\cdots\cdots \\ a_{n1}\xi_1 + a_{n2}\xi_2 + \cdots + a_{nn}\xi_n &= \lambda\xi_n,\end{aligned}$$

or

$$\begin{aligned}(a_{11} - \lambda)\xi_1 + a_{12}\xi_2 + \cdots + a_{1n}\xi_n &= 0, \\ a_{21}\xi_1 + (a_{22} - \lambda)\xi_2 + \cdots + a_{2n}\xi_n &= 0, \\ \cdots\cdots\cdots\cdots\cdots\cdots \\ a_{n1}\xi_1 + a_{n2}\xi_2 + \cdots + (a_{nn} - \lambda)\xi_n &= 0.\end{aligned} \tag{1}$$

Thus to prove the theorem we must show that there exists a number λ and a set of numbers $\xi_1, \xi_2, \cdots, \xi_n$ not all zero satisfying the system (1).

For the system (1) to have a non-trivial solution $\xi_1, \xi_2, \cdots, \xi_n$ it is necessary and sufficient that its determinant vanish, i.e., that

$$\begin{vmatrix} a_{11} - \lambda & a_{12} & \cdots & a_{1n} \\ a_{21} & a_{22} - \lambda & \cdots & a_{2n} \\ \cdots & \cdots & \cdots & \cdots \\ a_{n1} & a_{n2} & \cdots & a_{nn} - \lambda \end{vmatrix} = 0. \tag{2}$$

This polynomial equation of degree n in λ has at least one (in general complex) root λ_0.

With λ_0 in place of λ, (1) becomes a homogeneous system of linear equations with zero determinant. Such a system has a non-trivial solution $\xi_1^{(0)}, \xi_2^{(0)}, \cdots, \xi_n^{(0)}$. If we put

$$\mathbf{x}^{(0)} = \xi_1^{(0)}\mathbf{e}_1 + \xi_2^{(0)}\mathbf{e}_2 + \cdots + \xi_n^{(0)}\mathbf{e}_n,$$

then

$$\mathbf{Ax}^{(0)} = \lambda_0\mathbf{x}^{(0)},$$

i.e., $\mathbf{x}^{(0)}$ is an eigenvector and λ_0 an eigenvalue of A.

This completes the proof of the theorem.

NOTE: Since the proof remains valid when A is restricted to any subspace invariant under A, we can claim that *every invariant subspace contains at least one eigenvector of* A.

The polynomial on the left side of (2) is called the *characteristic polynomial* of the matrix of A and equation (2) the *characteristic equation* of that matrix. The proof of our theorem shows that the roots of the characteristic polynomial are eigenvalues of the transformation A and, conversely, the eigenvalues of A are roots of the characteristic polynomial.

Since the eigenvalues of a transformation are defined without reference to a basis, it follows that the roots of the characteristic polynomial do not depend on the choice of basis. In the sequel we shall prove a stronger result [2], namely, that the characteristic polynomial is itself independent of the choice of basis. We may thus speak of the characteristic polynomial of the transformation A rather than the characteristic polynomial of the matrix of the transformation A.

3. Linear transformations with n linearly independent eigenvectors are, in a way, the simplest linear transformations. Let A be such a transformation and $\mathbf{e}_1, \mathbf{e}_2, \cdots, \mathbf{e}_n$ its linearly independent eigenvectors, i.e.,

$$\mathrm{A}\mathbf{e}_i = \lambda_i \mathbf{e}_i \qquad (i = 1, 2, \cdots, n).$$

Relative to the basis $\mathbf{e}_1, \mathbf{e}_2, \cdots, \mathbf{e}_n$ the matrix of A is

$$\begin{bmatrix} \lambda_1 & 0 & \cdots & 0 \\ 0 & \lambda_2 & \cdots & 0 \\ 0 & 0 & \cdots & \lambda_n \end{bmatrix}.$$

Such a matrix is called a *diagonal matrix.* We thus have

THEOREM 2. *If a linear transformation* A *has n linearly independent eigenvectors then these vectors form a basis in which* A *is represented by a diagonal matrix. Conversely, if* A *is represented in some*

[2] The fact that the roots of the characteristic polynomial do not depend on the choice of basis does not by itself imply that the polynomial itself is independent of the choice of basis. It is a priori conceivable that the multiplicity of the roots varies with the basis.

basis by a diagonal matrix, then the vectors of this basis are eigenvalues of A.

NOTE: There is one important case in which a linear transformation is certain to have n linearly independent eigenvectors. We lead up to this case by observing that

If $\mathbf{e}_1, \mathbf{e}_2, \cdots, \mathbf{e}_k$ *are eigenvectors of a transformation* A *and the corresponding eigenvalues* $\lambda_1, \lambda_2, \cdots, \lambda_k$ *are distinct, then* $\mathbf{e}_1, \mathbf{e}_2, \cdots, \mathbf{e}_k$ *are linearly independent.*

For $k = 1$ this assertion is obviously true. We assume its validity for $k - 1$ vectors and prove it for the case of k vectors. If our assertion were false in the case of k vectors, then there would exist k numbers $\alpha_1, \alpha_2, \cdots, \alpha_k$, with $\alpha_1 \neq 0$, say, such that

$$\alpha_1 \mathbf{e}_1 + \alpha_2 \mathbf{e}_2 + \cdots + \alpha_k \mathbf{e}_k = \mathbf{0}. \tag{3}$$

Applying A to both sides of equation (3) we get

$$\mathrm{A}(\alpha_1 \mathbf{e}_1 + \alpha_2 \mathbf{e}_2 + \cdots + \alpha_k \mathbf{e}_k) = \mathbf{0},$$

or

$$\alpha_1 \lambda_1 \mathbf{e}_1 + \alpha_2 \lambda_2 \mathbf{e}_2 + \cdots + \alpha_k \lambda_k \mathbf{e}_k = \mathbf{0}.$$

Subtracting from this equation equation (3) multiplied by λ_k we are led to the relation

$$\alpha_1(\lambda_1 - \lambda_k)\mathbf{e}_1 + \alpha_2(\lambda_2 - \lambda_k)\mathbf{e}_2 + \cdots + \alpha_{k-1}(\lambda_{k-1} - \lambda_k)\mathbf{e}_{k-1} = \mathbf{0}$$

with $\lambda_1 - \lambda_k \neq 0$ (by assumption $\lambda_i \neq \lambda_k$ for $i \neq k$). This contradicts the assumed linear independence of $\mathbf{e}_1, \mathbf{e}_2, \cdots, \mathbf{e}_{k-1}$.

The following result is a direct consequence of our observation:

If the characteristic polynomial of a transformation A *has n distinct roots, then the matrix of* A *is diagonable.*

Indeed, a root λ_k of the characteristic equation determines at least one eigenvector. Since the λ_k are supposed distinct, it follows by the result just obtained that A has n linearly independent eigenvectors $\mathbf{e}_1, \mathbf{e}_2, \cdots, \mathbf{e}_n$. The matrix of A relative to the basis $\mathbf{e}_1, \mathbf{e}_2, \cdots, \mathbf{e}_n$ is diagonal.

If the characteristic polynomial has multiple roots, then the number of linearly independent eigenvectors may be less than n. For instance, the transformation A which associates with every polynomial of degree $\leq n - 1$ its derivative has only one eigenvalue $\lambda = 0$ and (to within a constant multiplier) one eigenvector $P(t) =$ constant. For if $P(t)$ is a polynomial of degree $k > 0$, then $P'(t)$ is a polynomial of degree $k - 1$. Hence $P'(t) = \lambda P(t)$ implies $\lambda = 0$ and $P(t) =$ constant, as asserted. It follows that regardless of the choice of basis the matrix of A is not diagonal.

We shall prove in chapter III that if λ is a root of multiplicity m of the characteristic polynomial of a transformation then the maximal number of linearly independent eigenvectors corresponding to λ is m.

In the sequel (§§ 12 and 13) we discuss a few classes of diagonable linear transformations (i.e., linear transformations which in some bases can be represented by diagonal matrices). The problem of the "simplest" matrix representation of an arbitrary linear transformation is discussed in chapter III.

4. *Characteristic polynomial.* In para. 2 we defined the characteristic polynomial of the matrix $\mathscr{A}$ of a linear transformation A as the determinant of the matrix $\mathscr{A} - \lambda\mathscr{E}$ and mentioned the fact that this polynomial is determined by the linear transformation A alone, i.e., it is independent of the choice of basis. In fact, if $\mathscr{A}$ and $\mathscr{B}$ represent A relative to two bases then $\mathscr{B} = \mathscr{C}^{-1}\mathscr{A}\mathscr{C}$ for some $\mathscr{C}$. But

$$|\mathscr{C}^{-1}\mathscr{A}\mathscr{C} - \lambda\mathscr{E}| = |\mathscr{C}^{-1}|\,|\mathscr{A} - \lambda\mathscr{E}|\,|\mathscr{C}| = |\mathscr{A} - \lambda\mathscr{E}|.$$

This proves our contention. Hence we can speak of the characteristic polynomial of a linear transformation (rather than the characteristic polynomial of the matrix of a linear transformation).

EXERCISES. *1.* Find the characteristic polynomial of the matrix

$$\begin{bmatrix} \lambda_0 & 0 & 0 & \cdots & 0 & 0 \\ 1 & \lambda_0 & 0 & \cdots & 0 & 0 \\ 0 & 1 & \lambda_0 & \cdots & 0 & 0 \\ \cdots & \cdots & \cdots & \cdots & \cdots & \cdots \\ 0 & 0 & 0 & \cdots & 1 & \lambda_0 \end{bmatrix}.$$

2. Find the characteristic polynomial of the matrix

$$\begin{bmatrix} a_1 & a_2 & a_3 & \cdots & a_{n-1} & a_n \\ 1 & 0 & 0 & \cdots & 0 & 0 \\ 0 & 1 & 0 & \cdots & 0 & 0 \\ \cdots & \cdots & \cdots & \cdots & \cdots & \cdots \\ 0 & 0 & 0 & \cdots & 1 & 0 \end{bmatrix}.$$

Solution: $(-1)^n(\lambda^n - a_1\lambda^{n-1} - a_2\lambda^{n-2} - \cdots - a_n)$.

We shall now find an explicit expression for the characteristic polynomial in terms of the entries in some representation $\mathscr{A}$ of A.

We begin by computing a more general polynomial, namely, $Q(\lambda) = |\mathscr{A} - \lambda\mathscr{B}|$, where $\mathscr{A}$ and $\mathscr{B}$ are two arbitrary matrices.

$$Q(\lambda) = \begin{vmatrix} a_{11} - \lambda b_{11} & a_{12} - \lambda b_{12} & \cdots & a_{1n} - \lambda b_{1n} \\ a_{21} - \lambda b_{21} & a_{22} - \lambda b_{22} & \cdots & a_{2n} - \lambda b_{2n} \\ \cdots & \cdots & \cdots & \cdots \\ a_{n1} - \lambda b_{n1} & a_{n2} - \lambda b_{n2} & \cdots & a_{nn} - \lambda b_{nn} \end{vmatrix}$$

and can (by the addition theorem on determinants) be written as the sum of determinants. The free term of $Q(\lambda)$ is

$$q_0 = \begin{vmatrix} a_{11} & a_{12} & \cdots & a_{1n} \\ a_{21} & a_{22} & \cdots & a_{2n} \\ \cdots & \cdots & \cdots & \cdots \\ a_{n1} & a_{n2} & \cdots & a_{nn} \end{vmatrix}. \tag{4}$$

The coefficient of $(-\lambda)^k$ *in the expression for* $Q(\lambda)$ *is the sum of determinants obtained by replacing in* (4) *any* k *columns of the matrix* $||a_{ik}||$ *by the corresponding columns of the matrix* $||b_{ik}||$.

In the case at hand $\mathscr{B} = \mathscr{E}$ and the determinants which add up to the coefficient of $(-\lambda^k)$ are the principal minors of order $n - k$ of the matrix $||a_{ik}||$. Thus, *the characteristic polynomial* $P(\lambda)$ *of the matrix* $\mathscr{A}$ *has the form*

$$P(\lambda) = (-1)^n(\lambda^n - p_1\lambda^{n-1} + p_2\lambda^{n-2} - \cdots \pm p_n),$$

where p_1 *is the sum of the diagonal entries of* $\mathscr{A}$, p_2 *the sum of the principal minors of order two, etc. Finally,* p_n *is the determinant of* $\mathscr{A}$.

We wish to emphasize the fact that the coefficients $p_1, p_2, \cdots, p_n$ are independent of the particular representation $\mathscr{A}$ of the transformation A. This is another way of saying that the characteristic polynomial is independent of the particular representation $\mathscr{A}$ of A.

The coefficients p_n and p_1 are of particular importance. p_n is the determinant of the matrix $\mathscr{A}$ and p_1 is the sum of the diagonal elements of $\mathscr{A}$. *The sum of the diagonal elements of* $\mathscr{A}$ *is called its trace.* It is clear that the trace of a matrix is the sum of all the roots of its characteristic polynomial each taken with its proper multiplicity.

To compute the eigenvectors of a linear transformation we must know its eigenvalues and this necessitates the solution of a polynomial equation of degree n. In one important case the roots of the characteristic polynomial can be read off from the matrix representing the transformation; namely,

If the matrix of a transformation A *is triangular, i.e., if it has the form*

$$(5)\qquad \begin{vmatrix} a_{11} & a_{12} & a_{13} & \cdots & a_{1n} \\ 0 & a_{22} & a_{23} & \cdots & a_{2n} \\ \cdots & \cdots & \cdots & \cdots & \cdots \\ 0 & 0 & 0 & \cdots & a_{nn} \end{vmatrix}$$

then the eigenvalues of A *are the numbers* $a_{11}, a_{22}, \cdots, a_{nn}$.

The proof is obvious since the characteristic polynomial of the matrix (5) is

$$P(\lambda) = (a_{11} - \lambda)(a_{22} - \lambda) \cdots (a_{nn} - \lambda)$$

and its roots are $a_{11}, a_{22}, \cdots, a_{nn}$.

EXERCISE. Find the eigenvectors corresponding to the eigenvalues a_{11}, a_{22}, a_{33} of the matrix (5).

We conclude with a discussion of an interesting property of the characteristic polynomial. As was pointed out in para. 3 of § 9, for every matrix $\mathscr{A}$ there exists a polynomial $P(t)$ such that $P(\mathscr{A})$ is the zero matrix. We now show that the characteristic polynomial is just such a polynomial. First we prove the following

LEMMA 1. *Let the polynomial*

$$P(\lambda) = a_0\lambda^m + a_1\lambda^{m-1} + \cdots + a_m$$

and the matrix $\mathscr{A}$ *be connected by the relation*

$$(6)\qquad P(\lambda)\mathscr{E} = (\mathscr{A} - \lambda\mathscr{E})\mathscr{C}(\lambda)$$

where $\mathscr{C}(\lambda)$ *is a polynomial in* λ *with matrix coefficients, i.e.,*

$$\mathscr{C}(\lambda) = \mathscr{C}_0\lambda^{m-1} + \mathscr{C}_1\lambda^{m-2} + \cdots + \mathscr{C}_{m-1}.$$

Then $P(\mathscr{A}) = \mathscr{O}$.

(We note that this lemma is an extension of the theorem of Bezout to polynomials with matrix coefficients.)

Proof: We have

$$(7)\qquad (\mathscr{A} - \lambda\mathscr{E})\mathscr{C}(\lambda) = \mathscr{A}\mathscr{C}_{m-1} + (\mathscr{A}\mathscr{C}_{m-2} - \mathscr{C}_{m-1})\lambda + (\mathscr{A}\mathscr{C}_{m-3} - \mathscr{C}_{m-2})\lambda^2 + \cdots - \mathscr{C}_0\lambda^m.$$

Now (6) and (7) yield the equations

$$(8)\qquad \begin{aligned} \mathscr{A}\mathscr{C}_{m-1} &= a_m\mathscr{E}, \\ \mathscr{A}\mathscr{C}_{m-2} - \mathscr{C}_{m-1} &= a_{m-1}\mathscr{E}, \\ \mathscr{A}\mathscr{C}_{m-3} - \mathscr{C}_{m-2} &= a_{m-2}\mathscr{E}, \\ &\cdots\cdots\cdots \\ \mathscr{A}\mathscr{C}_0 - \mathscr{C}_1 &= a_1\mathscr{E}, \\ -\mathscr{C}_0 &= a_0\mathscr{E}. \end{aligned}$$

If we multiply the first of these equations on the left by $\mathscr{E}$, the second by $\mathscr{A}$, the third by $\mathscr{A}^2, \cdots$, the last by $\mathscr{A}^m$ and add the resulting equations, we get $\mathscr{O}$ on the left, and $P(\mathscr{A}) = a_m \mathscr{E} + a_{m-1} \mathscr{A} + \cdots + a_0 \mathscr{A}^m$ on the right. Thus $P(\mathscr{A}) = \mathscr{O}$ and our lemma is proved [3].

THEOREM 3. *If $P(\lambda)$ is the characteristic polynomial of $\mathscr{A}$, then $P(\mathscr{A}) = \mathscr{O}$.*

Proof: Consider the inverse of the matrix $\mathscr{A} - \lambda\mathscr{E}$. We have $(\mathscr{A} - \lambda\mathscr{E})(\mathscr{A} - \lambda\mathscr{E})^{-1} = \mathscr{E}$. As is well known, the inverse matrix can be written in the form

$$(\mathscr{A} - \lambda\mathscr{E})^{-1} = \frac{1}{P(\lambda)} \mathscr{C}(\lambda),$$

where $\mathscr{C}(\lambda)$ is the matrix of the cofactors of the elements of $\mathscr{A} - \lambda\mathscr{E}$ and $P(\lambda)$ the determinant of $\mathscr{A} - \lambda\mathscr{E}$, i.e., the characteristic polynomial of $\mathscr{A}$. Hence

$$(\mathscr{A} - \lambda\mathscr{E})\mathscr{C}(\lambda) = P(\lambda)\mathscr{E}.$$

Since the elements of $\mathscr{C}(\lambda)$ are polynomials of degree $\leqq n - 1$ in λ, we conclude on the basis of our lemma that

$$P(\mathscr{A}) = \mathscr{O}.$$

This completes the proof.

We note that if the characteristic polynomial of the matrix $\mathscr{A}$ has no multiple roots, then there exists no polynomial $Q(\lambda)$ of degree less than n such that $Q(\mathscr{A}) = \mathscr{O}$ (cf. the exercise below).

EXERCISE. Let $\mathscr{A}$ be a diagonal matrix

$$\mathscr{A} = \begin{bmatrix} \lambda_1 & 0 & \cdots & 0 \\ 0 & \lambda_2 & \cdots & 0 \\ \cdots & \cdots & \cdots & \cdots \\ 0 & 0 & \cdots & \lambda_n \end{bmatrix},$$

where all the λ_i are distinct. Find a polynomial $P(t)$ of lowest degree for which $P(\mathscr{A}) = \mathscr{O}$ (cf. para. 3, § 9).

§ 11. The adjoint of a linear transformation

1. *Connection between transformations and bilinear forms in Euclidean space.* We have considered under separate headings linear transformations and bilinear forms on vector spaces. In

[3] In algebra the theorem of Bezout is proved by direct substitution of A in (6). Here this is not an admissible procedure since λ is a number and $\mathscr{A}$ is a matrix. However, we are doing essentially the same thing. In fact, the kth equation in (8) is obtained by equating the coefficients of λ^k in (6). Subsequent multiplication by $\mathscr{A}^k$ and addition of the resulting equations is tantamount to the substitution of $\mathscr{A}$ in place of λ.

the case of Euclidean spaces there exists a close connection between bilinear forms and linear transformations [4].

Let **R** be a complex Euclidean space and let $A(\mathbf{x}; \mathbf{y})$ be a bilinear form on **R**. Let $\mathbf{e}_1, \mathbf{e}_2, \cdots, \mathbf{e}_n$ be an orthonormal basis in **R**. If $\mathbf{x} = \xi_1\mathbf{e}_1 + \xi_2\mathbf{e}_2 + \cdots + \xi_n\mathbf{e}_n$ and $\mathbf{y} = \eta_1\mathbf{e}_1 + \eta_2\mathbf{e}_2 + \cdots + \eta_n\mathbf{e}_n$, then $A(\mathbf{x}; \mathbf{y})$ can be written in the form

$$\begin{aligned} A(\mathbf{x}; \mathbf{y}) = {} & a_{11}\xi_1\bar{\eta}_1 + a_{12}\xi_1\bar{\eta}_2 + \cdots + a_{1n}\xi_1\bar{\eta}_n \\ & + a_{21}\xi_2\bar{\eta}_1 + a_{22}\xi_2\bar{\eta}_2 + \cdots + a_{2n}\xi_2\bar{\eta}_n \\ & + \cdots\cdots\cdots\cdots\cdots\cdots \\ & + a_{n1}\xi_n\bar{\eta}_1 + a_{n2}\xi_n\bar{\eta}_2 + \cdots + a_{nn}\xi_n\bar{\eta}_n. \end{aligned} \tag{1}$$

We shall now try to represent the above expression as an inner product. To this end we rewrite it as follows:

$$\begin{aligned} A(\mathbf{x}; \mathbf{y}) = {} & (a_{11}\xi_1 + a_{21}\xi_2 + \cdots + a_{n1}\xi_n)\bar{\eta}_1 \\ & + (a_{12}\xi_1 + a_{22}\xi_2 + \cdots + a_{n2}\xi_n)\bar{\eta}_2 \\ & + \cdots\cdots\cdots\cdots\cdots\cdots \\ & + (a_{1n}\xi_1 + a_{2n}\xi_2 + \cdots + a_{nn}\xi_n)\bar{\eta}_n. \end{aligned}$$

Now we introduce the vector $\mathbf{z}$ with coordinates

$$\begin{aligned} \zeta_1 &= a_{11}\xi_1 + a_{21}\xi_2 + \cdots + a_{n1}\xi_n, \\ \zeta_2 &= a_{12}\xi_1 + a_{22}\xi_2 + \cdots + a_{n2}\xi_n, \\ & \cdots\cdots\cdots\cdots\cdots\cdots \\ \zeta_n &= a_{1n}\xi_1 + a_{2n}\xi_2 + \cdots + a_{nn}\xi_n. \end{aligned}$$

It is clear that $\mathbf{z}$ is obtained by applying to $\mathbf{x}$ a linear transformation whose matrix is the transpose of the matrix $||a_{ik}||$ of the bilinear form $A(\mathbf{x}; \mathbf{y})$. We shall denote this linear transformation

[4] Relative to a given basis both linear transformations and bilinear forms are given by matrices. One could therefore try to associate with a given linear transformation the bilinear form determined by the same matrix as the transformation in question. However, such correspondence would be without significance. In fact, if a linear transformation and a bilinear form are represented relative to some basis by a matrix $\mathscr{A}$, then, upon change of basis, the linear transformation is represented by $\mathscr{C}^{-1}\mathscr{A}\mathscr{C}$ (cf. § 9) and the bilinear form is represented by $\mathscr{C}'\mathscr{A}\mathscr{C}$ (cf. § 4). Here $\mathscr{C}'$ is the transpose of $\mathscr{C}$.

The careful reader will notice that the correspondence between bilinear forms and linear transformations in Euclidean space considered below associates bilinear forms and linear transformations whose matrices relative to an orthonormal basis are transposes of one another. This correspondence is shown to be independent of the choice of basis.

by the letter A, i.e., we shall put $\mathbf{z} = \mathrm{A}\mathbf{x}$. Then

$$A(\mathbf{x}; \mathbf{y}) = \zeta_1\bar{\eta}_1 + \zeta_2\bar{\eta}_2 + \cdots + \zeta_n\bar{\eta}_n = (\mathbf{z}, \mathbf{y}) = (\mathrm{A}\mathbf{x}, \mathbf{y}).$$

Thus, *a bilinear form* $A(\mathbf{x}; \mathbf{y})$ *on Euclidean vector space determines a linear transformation* A *such that*

$$A(\mathbf{x}; \mathbf{y}) \equiv (\mathrm{A}\mathbf{x}, \mathbf{y}).$$

The converse of this proposition is also true, namely:

A linear transformation A *on a Euclidean vector space determines a bilinear form* $A(\mathbf{x}; \mathbf{y})$ *defined by the relation*

$$A(\mathbf{x}; \mathbf{y}) \equiv (\mathrm{A}\mathbf{x}, \mathbf{y}).$$

The bilinearity of $A(\mathbf{x}; \mathbf{y}) \equiv (\mathrm{A}\mathbf{x}, \mathbf{y})$ is easily proved:

1. $(\mathrm{A}(\mathbf{x}_1 + \mathbf{x}_2), \mathbf{y}) = (\mathrm{A}\mathbf{x}_1 + \mathrm{A}\mathbf{x}_2, \mathbf{y}) = (\mathrm{A}\mathbf{x}_1, \mathbf{y}) + (\mathrm{A}\mathbf{x}_2, \mathbf{y})$,
   $(\mathrm{A}\lambda\mathbf{x}, \mathbf{y}) = (\lambda\mathrm{A}\mathbf{x}, \mathbf{y}) = \lambda(\mathrm{A}\mathbf{x}, \mathbf{y})$.
2. $(\mathbf{x}, \mathrm{A}(\mathbf{y}_1 + \mathbf{y}_2)) = (\mathbf{x}, \mathrm{A}\mathbf{y}_1 + \mathrm{A}\mathbf{y}_2) = (\mathbf{x}, \mathrm{A}\mathbf{y}_1) + (\mathbf{x}, \mathrm{A}\mathbf{y}_2)$,
   $(\mathbf{x}, \mathrm{A}\mu\mathbf{y}) = (\mathbf{x}, \mu\mathrm{A}\mathbf{y}) = \bar{\mu}(\mathbf{x}, \mathrm{A}\mathbf{y})$.

We now show that the bilinear form $A(\mathbf{x}; \mathbf{y})$ determines the transformation A uniquely. Thus, let

$$A(\mathbf{x}; \mathbf{y}) = (\mathrm{A}\mathbf{x}, \mathbf{y})$$

and

$$A(\mathbf{x}; \mathbf{y}) = (\mathrm{B}\mathbf{x}, \mathbf{y}).$$

Then

$$(\mathrm{A}\mathbf{x}, \mathbf{y}) \equiv (\mathrm{B}\mathbf{x}, \mathbf{y}),$$

i.e.,

$$(\mathrm{A}\mathbf{x} - \mathrm{B}\mathbf{x}, \mathbf{y}) = 0$$

for all $\mathbf{y}$. But this means that $\mathrm{A}\mathbf{x} - \mathrm{B}\mathbf{x} = \mathbf{0}$ for all $\mathbf{x}$. Hence $\mathrm{A}\mathbf{x} = \mathrm{B}\mathbf{x}$ for all $\mathbf{x}$, which is the same as saying that $\mathrm{A} = \mathrm{B}$. This proves the uniqueness assertion.

We can now sum up our results in the following

THEOREM 1. *The equation*

$$A(\mathbf{x}; \mathbf{y}) = (\mathrm{A}\mathbf{x}, \mathbf{y}) \tag{2}$$

establishes a one-to-one correspondence between bilinear forms and linear transformations on a Euclidean vector space.

The one-oneness of the correspondence established by eq. (2) implies its independence from choice of basis.

There is another way of establishing a connection between bilinear forms and linear transformations. Namely, every bilinear form can be represented as

$$A(\mathbf{x}; \mathbf{y}) = (\mathbf{x}, \mathrm{A}^*\mathbf{y}).$$

This representation is obtained by rewriting formula (1) above in the following manner:

$$\begin{aligned} A(\mathbf{x}; \mathbf{y}) &= \xi_1(a_{11}\bar{\eta}_1 + a_{12}\bar{\eta}_2 + \cdots + a_{1n}\bar{\eta}_n) \\ &+ \xi_2(a_{21}\bar{\eta}_1 + a_{22}\bar{\eta}_2 + \cdots + a_{2n}\bar{\eta}_n) \\ &+ \cdots\cdots\cdots\cdots\cdots\cdots \\ &+ \xi_n(a_{n1}\bar{\eta}_1 + a_{n2}\bar{\eta}_2 + \cdots + a_{nn}\bar{\eta}_n) \\ &= \xi_1(\overline{\bar{a}_{11}\eta_1 + \bar{a}_{12}\eta_2 + \cdots + \bar{a}_{1n}\eta_n}) \\ &+ \xi_2(\overline{\bar{a}_{21}\eta_1 + \bar{a}_{22}\eta_2 + \cdots + \bar{a}_{2n}\eta_n}) \\ &+ \cdots\cdots\cdots\cdots\cdots\cdots \\ &+ \xi_n(\overline{\bar{a}_{n1}\eta_1 + \bar{a}_{n2}\eta_2 + \cdots + \bar{a}_{nn}\eta_n}) = (\mathbf{x}, \mathrm{A}^*\mathbf{y}). \end{aligned}$$

Relative to an *orthogonal* basis the matrix $||a^*_{ik}||$ of A* and the matrix $||a_{ik}||$ of A are connected by the relation

$$a^*_{ik} = \bar{a}_{ki}.$$

For a non-orthogonal basis the connection between the two matrices is more complicated.

2. *Transition from* A *to its adjoint* (*the operation* *)

DEFINITION 1. *Let* A *be a linear transformation on a complex Euclidean space. The transformation* A* *defined by*

$$(\mathrm{A}\mathbf{x}, \mathbf{y}) = (\mathbf{x}, \mathrm{A}^*\mathbf{y})$$

is called the adjoint of A.

THEOREM 2. *In a Euclidean space there is a one-to-one correspondence between linear transformations and their adjoints.*

Proof: According to Theorem 1 of this section every linear transformation determines a unique bilinear form $A(\mathbf{x}; \mathbf{y}) = (\mathrm{A}\mathbf{x}, \mathbf{y})$. On the other hand, by the result stated in the conclusion of para. 1, every bilinear form can be uniquely represented as $(\mathbf{x}, \mathrm{A}^*\mathbf{y})$. Hence

$$(\mathbf{Ax}, \mathbf{y}) = A(\mathbf{x}; \mathbf{y}) = (\mathbf{x}, \mathbf{A}^*\mathbf{y}).$$

The connection between the matrices of A and A* relative to an *orthogonal* matrix was discussed above.

Some of the basic properties of the operation * are

1. $(\mathrm{AB})^* = \mathrm{B}^*\mathrm{A}^*$.
2. $(\mathrm{A}^*)^* = \mathrm{A}$.
3. $(\mathrm{A} + \mathrm{B})^* = \mathrm{A}^* + \mathrm{B}^*$.
4. $(\lambda\mathrm{A})^* = \bar{\lambda}\mathrm{A}^*$.
5. $\mathrm{E}^* = \mathrm{E}$.

We give proofs of properties 1 and 2.

1. $(\mathrm{AB}\mathbf{x}, \mathbf{y}) = (\mathrm{B}\mathbf{x}, \mathrm{A}^*\mathbf{y}) = (\mathbf{x}, \mathrm{B}^*\mathrm{A}^*\mathbf{y})$.

On the other hand, the definition of (AB)* implies

$$(\mathrm{AB}\mathbf{x}, \mathbf{y}) = (\mathbf{x}, (\mathrm{AB})^*\mathbf{y}).$$

If we compare the right sides of the last two equations and recall that a linear transformation is uniquely determined by the corresponding bilinear form we conclude that

$$(\mathrm{AB})^* = \mathrm{B}^*\mathrm{A}^*.$$

2. By the definition of A*,

$$(\mathrm{A}\mathbf{x}, \mathbf{y}) = (\mathbf{x}, \mathrm{A}^*\mathbf{y}).$$

Denote A* by C. Then

$$(\mathrm{A}\mathbf{x}, \mathbf{y}) = (\mathbf{x}, \mathrm{C}\mathbf{y}),$$

whence

$$(\mathbf{y}, \mathrm{A}\mathbf{x}) = (\mathrm{C}\mathbf{y}, \mathbf{x}).$$

Interchange of **x** and **y** gives

$$(\mathrm{C}\mathbf{x}, \mathbf{y}) = (\mathbf{x}, \mathrm{A}\mathbf{y}).$$

But this means that $\mathrm{C}^* = \mathrm{A}$, i.e., $(\mathrm{A}^*)^* = \mathrm{A}$.

EXERCISES. *1.* Prove properties 3 through 5 of the operation *.

2. Prove properties 1 through 5 of the operation * by making use of the connection between the matrices of A and A* relative to an orthogonal basis.

3. *Self-adjoint, unitary and normal linear transformations.* The operation * is to some extent the analog of the operation of

conjugation which takes a complex number α into the complex number $\bar{\alpha}$. This analogy is not accidental. Indeed, it is clear that for matrices of order one over the field of complex numbers, i.e., for complex numbers, the two operations are the same.

The real numbers are those complex numbers for which $\bar{\alpha} = \alpha$. The class of linear transformations which are the analogs of the real numbers is of great importance. This class is introduced by

DEFINITION 2. *A linear transformation is called self-adjoint (Hermitian) if* $A^* = A$.

We now show that *for a linear transformation* A *to be self-adjoint it is necessary and sufficient that the bilinear form* $(A\mathbf{x}, \mathbf{y})$ *be Hermitian.*

Indeed, to say that the form $(A\mathbf{x}, \mathbf{y})$ is Hermitian is to say that

$$\text{(a)} \qquad (A\mathbf{x}, \mathbf{y}) = \overline{(A\mathbf{y}, \mathbf{x})}.$$

Again, to say that A is self-adjoint is to say that

$$\text{(b)} \qquad (A\mathbf{x}, \mathbf{y}) = (\mathbf{x}, A\mathbf{y}).$$

Clearly, equations (a) and (b) are equivalent.

Every complex number ζ is representable in the form $\zeta = \alpha + i\beta$, α, β real. Similarly,

Every linear transformation A *can be written as a sum*

$$(3) \qquad A = A_1 + iA_2,$$

where A_1 *and* A_2 *are self-adjoint transformations.*

In fact, let $A_1 = (A + A^*)/2$ and $A_2 = (A - A^*)/2i$. Then $A = A_1 + iA_2$ and

$$A_1^* = \left(\frac{A + A^*}{2}\right)^* = \tfrac{1}{2}(A + A^*)^* = \tfrac{1}{2}(A^* + A^{**}) = \tfrac{1}{2}(A^* + A) = A_1,$$

$$A_2^* = \left(\frac{A - A^*}{2i}\right)^* = -\frac{1}{2i}(A - A^*)^* = -\frac{1}{2i}(A^* - A^{**}) = -\frac{1}{2i}(A^* - A) = A_2,$$

i.e., A_1 and A_2 are self-adjoint.

This brings out the analogy between real numbers and self-adjoint transformations.

EXERCISES. *1.* Prove the uniqueness of the representation (3) of A.

2. Prove that a linear combination with real coefficients of self-adjoint transformations is again self-adjoint.

3. Prove that if A is an arbitrary linear transformation then AA^* and A^*A are self-adjoint.

NOTE: In contradistinction to complex numbers AA^* is, in general, different from A^*A.

The product of two self-adjoint transformations is, in general, not self-adjoint. However:

THEOREM 3. *For the product* AB *of two self-adjoint transformations* A *and* B *to be self-adjoint it is necessary and sufficient that* A *and* B *commute.*

Proof: We know that

$$A^* = A \quad \text{and} \quad B^* = B.$$

We wish to find a condition which is necessary and sufficient for

$$(AB)^* = AB. \tag{4}$$

Now,

$$(AB)^* = B^*A^* = BA.$$

Hence (4) is equivalent to the equation

$$AB = BA.$$

This proves the theorem.

EXERCISE. Show that if A and B are self-adjoint, then $AB + BA$ and $i(AB - BA)$ are also self-adjoint.

The analog of complex numbers of absolute value one are unitary transformations.

DEFINITION 3. *A linear transformation* U *is called unitary if* $UU^* = U^*U = E$.[5] In other words for a unitary transformations $U^* = U^{-1}$.

In § 13 we shall become familiar with a very simple geometric interpretation of unitary transformations.

EXERCISES. *1.* Show that the product of two unitary transformations is a unitary transformation.

2. Show that if U is unitary and A self-adjoint, then $U^{-1}AU$ is again self-adjoint.

[5] In *n*-dimensional spaces $UU^* = E$ and $U^*U = E$ are equivalent statements. This is not the case in infinite dimensional spaces.

In the sequel (§ 15) we shall prove that every linear transformation can be written as the product of a self-adjoint transformation and a unitary transformation. This result can be regarded as a generalization of the result on the trigonometric form of a complex number.

DEFINITION 4. *A linear transformation* A *is called normal if* $AA^* = A^*A$.

There is no need to introduce an analogous concept in the field of complex numbers since multiplication of complex numbers is commutative.

It is easy to see that unitary transformations and self-adjoint transformations are normal.

The subsequent sections of this chapter are devoted to a more detailed study of the various classes of linear transformations just introduced. In the course of this study we shall become familiar with very simple geometric characterizations of these classes of transformations.

§ 12. Self-adjoint (Hermitian) transformations. Simultaneous reduction of a pair of quadratic forms to a sum of squares

1. *Self-adjoint transformations.* This section is devoted to a more detailed study of self-adjoint transformations on n-dimensional Euclidean space. These transformations are frequently encountered in different applications. (Self-adjoint transformations on infinite dimensional space play an important role in quantum mechanics.)

LEMMA 1. *The eigenvalues of a self-adjoint transformation are real.*

Proof: Let $\mathbf{x}$ be an eigenvector of a self-adjoint transformation A and let λ be the eigenvalue corresponding to $\mathbf{x}$, i.e.,

$$A\mathbf{x} = \lambda\mathbf{x}; \qquad \mathbf{x} \neq \mathbf{0}.$$

Since $A^* = A$,

$$(A\mathbf{x}, \mathbf{x}) = (\mathbf{x}, A\mathbf{x}),$$

that is,

$$(\lambda\mathbf{x}, \mathbf{x}) = (\mathbf{x}, \lambda\mathbf{x}),$$

or,

$$\lambda(\mathbf{x}, \mathbf{x}) = \bar{\lambda}(\mathbf{x}, \mathbf{x}).$$

Since $(\mathbf{x}, \mathbf{x}) \neq 0$, it follows that $\lambda = \bar{\lambda}$, which proves that λ is real.

LEMMA 2. *Let* A *be a self-adjoint transformation on an n-dimensional Euclidean vector space* **R** *and let* **e** *be an eigenvector of* A. *The totality* $\mathbf{R}_1$ *of vectors* **x** *orthogonal to* **e** *form an* $(n - 1)$*-dimensional subspace invariant under* A.

Proof: The totality $\mathbf{R}_1$ of vectors **x** orthogonal to **e** form an $(n - 1)$-dimensional subspace of **R**.

We show that $\mathbf{R}_1$ is invariant under A. Let $\mathbf{x} \in \mathbf{R}_1$. This means that $(\mathbf{x}, \mathbf{e}) = 0$. We have to show that $A\mathbf{x} \in \mathbf{R}_1$, that is, $(A\mathbf{x}, \mathbf{e}) = 0$. Indeed,

$$(A\mathbf{x}, \mathbf{e}) = (\mathbf{x}, A^*\mathbf{e}) = (\mathbf{x}, A\mathbf{e}) = (\mathbf{x}, \lambda\mathbf{e}) = \lambda(\mathbf{x}, \mathbf{e}) = 0.$$

THEOREM 1. *Let* A *be a self-adjoint transformation on an n-dimensional Euclidean space. Then there exist n pairwise orthogonal eigenvectors of* A. *The corresponding eigenvalues of* A *are all real.*

Proof: According to Theorem 1, § 10, there exists at least one eigenvector $\mathbf{e}_1$ of A. By Lemma 2, the totality of vectors orthogonal to $\mathbf{e}_1$ form an $(n - 1)$-dimensional invariant subspace $\mathbf{R}_1$. We now consider our transformation A on $\mathbf{R}_1$ only. In $\mathbf{R}_1$ there exists a vector $\mathbf{e}_2$ which is an eigenvector of A (cf. note to Theorem 1, § 10). The totality of vectors of $\mathbf{R}_1$ orthogonal to $\mathbf{e}_2$ form an $(n - 2)$-dimensional invariant subspace $\mathbf{R}_2$. In $\mathbf{R}_2$ there exists an eigenvector $\mathbf{e}_3$ of A, etc.

In this manner we obtain n pairwise orthogonal eigenvectors $\mathbf{e}_1, \mathbf{e}_2, \cdots, \mathbf{e}_n$. By Lemma 1, the corresponding eigenvalues are real. This proves Theorem 1.

Since the product of an eigenvector by any non-zero number is again an eigenvector, we can select the vectors $\mathbf{e}_i$ so that each of them is of length one.

THEOREM 2. *Let* A *be a linear transformation on an n-dimensional Euclidean space* **R**. *For* A *to be self-adjoint it is necessary and sufficient that there exists an orthogonal basis relative to which the matrix of* A *is diagonal and real.*

Necessity: Let A be self-adjoint. Select in **R** a basis consisting of the n pairwise orthogonal eigenvectors $\mathbf{e}_1, \mathbf{e}_2, \cdots, \mathbf{e}_n$ of A constructed in the proof of Theorem 1.

Since

$$\begin{aligned} A\mathbf{e}_1 &= \lambda_1 \mathbf{e}_1, \\ A\mathbf{e}_2 &= \lambda_2 \mathbf{e}_2, \\ &\cdots\cdots\cdots \\ A\mathbf{e}_n &= \lambda_n \mathbf{e}_n, \end{aligned}$$

it follows that relative to this basis the matrix of the transformation A is of the form

$$\begin{bmatrix} \lambda_1 & 0 & \cdots & 0 \\ 0 & \lambda_2 & \cdots & 0 \\ \cdots & \cdots & \cdots & \cdots \\ 0 & 0 & \cdots & \lambda_n \end{bmatrix} \tag{1}$$

where the λ_i are real.

Sufficiency: Assume now that the matrix of the transformation A has relative to an orthogonal basis the form (1). The matrix of the adjoint transformation A* relative to an orthonormal basis is obtained by replacing all entries in the transpose of the matrix of A by their conjugates (cf. § 11). In our case this operation has no effect on the matrix in question. Hence the transformations A and A* have the same matrix, i.e., A = A*. This concludes the proof of Theorem 2.

We note the following property of the eigenvectors of a self-adjoint transformation: *the eigenvectors corresponding to different eigenvalues are orthogonal.*

Indeed, let

$$A\mathbf{e}_1 = \lambda_1 \mathbf{e}_1, \qquad A\mathbf{e}_2 = \lambda_2 \mathbf{e}_2, \qquad \lambda_1 \neq \lambda_2.$$

Then

$$(A\mathbf{e}_1, \mathbf{e}_2) = (\mathbf{e}_1, A^*\mathbf{e}_2) = (\mathbf{e}_1, A\mathbf{e}_2),$$

that is

$$\lambda_1(\mathbf{e}_1, \mathbf{e}_2) = \lambda_2(\mathbf{e}_1, \mathbf{e}_2),$$

or

$$(\lambda_1 - \lambda_2)(\mathbf{e}_1, \mathbf{e}_2) = 0.$$

Since $\lambda_1 \neq \lambda_2$, it follows that

$$(\mathbf{e}_1, \mathbf{e}_2) = 0.$$

NOTE: Theorem 2 suggests the following geometric interpretation of a self-adjoint transformation: We select in our space n pairwise orthogonal directions (the directions determined by the eigenvectors) and associate with each a real number λ_i (eigenvalue). Along each one of these directions we perform a stretching by $|\lambda_i|$ and, in addition, if λ_i happens to be negative, a reflection in the plane orthogonal to the corresponding direction.

Along with the notion of a self-adjoint transformation we introduce the notion of a Hermitian matrix.

The matrix $||a_{ik}||$ is said to be *Hermitian* if $a_{ik} = \bar{a}_{ki}$.

Clearly, a necessary and sufficient condition for a linear transformation A to be self-adjoint is that its matrix relative to some orthogonal basis be Hermitian.

EXERCISE. Raise the matrix

$$\begin{pmatrix} 0 & \sqrt{2} \\ \sqrt{2} & 1 \end{pmatrix}$$

to the 28th power. *Hint*: Bring the matrix to its diagonal form, raise it to the proper power, and then revert to the original basis.

2. *Reduction to principal axes. Simultaneous reduction of a pair of quadratic forms to a sum of squares.* We now apply the results obtained in para. 1 to quadratic forms.

We know that we can associate with each Hermitian bilinear form a self-adjoint transformation. Theorem 2 permits us now to state the important

THEOREM 3. *Let $A(\mathbf{x}; \mathbf{y})$ be a Hermitian bilinear form defined on an n-dimensional Euclidean space* **R**. *Then there exists an orthonormal basis in* **R** *relative to which the corresponding quadratic form can be written as a sum of squares,*

$$A(\mathbf{x}; \mathbf{x}) = \sum \lambda_i |\xi_i|^2,$$

where the λ_i are real, and the ξ_i are the coordinates of the vector **x**.[6]

Proof: Let $A(\mathbf{x}; \mathbf{y})$ be a Hermitian bilinear form, i.e.,

$$A(\mathbf{x}; \mathbf{y}) = \overline{A(\mathbf{y}; \mathbf{x})},$$

[6] We have shown in § 8 that in any vector space a Hermitian quadratic form can be written in an appropriate basis as a sum of squares. In the case of a *Euclidean* space we can state a stronger result, namely, we can assert the existence of an *orthonormal basis* relative to which a given Hermitian quadratic form can be reduced to a sum of squares.

then there exists (cf. § 11) a self-adjoint linear transformation A such that

$$A(\mathbf{x}; \mathbf{y}) \equiv (\mathrm{A}\mathbf{x}, \mathbf{y}).$$

As our orthonormal basis vectors we select the pairwise orthogonal eigenvectors $\mathbf{e}_1, \mathbf{e}_2, \cdots, \mathbf{e}_n$ of the self-adjoint transformation A (cf. Theorem 1). Then

$$\mathrm{A}\mathbf{e}_1 = \lambda_1 \mathbf{e}_1, \qquad \mathrm{A}\mathbf{e}_2 = \lambda_2 \mathbf{e}_2, \qquad \cdots, \qquad \mathrm{A}\mathbf{e}_n = \lambda_n \mathbf{e}_n.$$

Let

$$\mathbf{x} = \xi_1 \mathbf{e}_1 + \xi_2 \mathbf{e}_2 + \cdots + \xi_n \mathbf{e}_n, \qquad \mathbf{y} = \eta_1 \mathbf{e}_1 + \eta_2 \mathbf{e}_2 + \cdots + \eta_n \mathbf{e}_n.$$

Since

$$(\mathbf{e}_i, \mathbf{e}_k) = \begin{cases} 1 & \text{for } i = k \\ 0 & \text{for } i \neq k, \end{cases}$$

we get

$$\begin{aligned} A(\mathbf{x}; \mathbf{y}) &\equiv (\mathrm{A}\mathbf{x}, \mathbf{y}) \\ &= (\xi_1 \mathrm{A}\mathbf{e}_1 + \xi_2 \mathrm{A}\mathbf{e}_2 + \cdots + \xi_n \mathrm{A}\mathbf{e}_n, \eta_1 \mathbf{e}_1 + \eta_2 \mathbf{e}_2 + \cdots + \eta_n \mathbf{e}_n) \\ &= (\lambda_1 \xi_1 \mathbf{e}_1 + \lambda_2 \xi_2 \mathbf{e}_2 + \cdots + \lambda_n \xi_n \mathbf{e}_n, \eta_1 \mathbf{e}_1 + \eta_2 \mathbf{e}_2 + \cdots + \eta_n \mathbf{e}_n) \\ &= \lambda_1 \xi_1 \bar{\eta}_1 + \lambda_2 \xi_2 \bar{\eta}_2 + \cdots + \lambda_n \xi_n \bar{\eta}_n. \end{aligned}$$

In particular,

$$A(\mathbf{x}; \mathbf{x}) = (\mathrm{A}\mathbf{x}, \mathbf{x}) = \lambda_1 |\xi_1|^2 + \lambda_2 |\xi_2|^2 + \cdots + \lambda_n |\xi_n|^2.$$

This proves the theorem.

The process of finding an orthonormal basis in a Euclidean space relative to which a given quadratic form can be represented as a sum of squares is called reduction to principal axes.

THEOREM 4. *Let $A(\mathbf{x}; \mathbf{x})$ and $B(\mathbf{x}; \mathbf{x})$ be two Hermitian quadratic forms on an n-dimensional vector space* **R** *and assume $B(\mathbf{x}; \mathbf{x})$ to be positive definite. Then there exists a basis in* **R** *relative to which each form can be written as a sum of squares.*

Proof: We introduce in **R** an inner product by putting $(\mathbf{x}, \mathbf{y}) \equiv B(\mathbf{x}; \mathbf{y})$, where $B(\mathbf{x}; \mathbf{y})$ is the bilinear form corresponding to $B(\mathbf{x}; \mathbf{x})$. This can be done since the axioms for an inner product state that $(\mathbf{x}, \mathbf{y})$ is a Hermitian bilinear form corresponding to a positive definite quadratic form (§ 8). With the introduction of an inner product our space **R** becomes a Euclidean vector space. By theorem 3 $\mathbf{R}$ contains an orthonormal[7] basis $\mathbf{e}_1, \mathbf{e}_2, \cdots, \mathbf{e}_n$ relative to which the form $A(\mathbf{x}; \mathbf{x})$ can be written as a sum of squares,

$$(2) \qquad A(\mathbf{x}; \mathbf{x}) = \lambda_1|\xi_1|^2 + \lambda_2|\xi_2|^2 + \cdots + \lambda_n|\xi_n|^2.$$

Now, with respect to an orthonormal basis an inner product takes the form

$$(\mathbf{x}, \mathbf{x}) = |\xi_1|^2 + |\xi_2|^2 + \cdots + |\xi_n|^2.$$

Since $B(\mathbf{x}; \mathbf{x}) \equiv (\mathbf{x}, \mathbf{x})$, it follows that

$$(3) \qquad B(\mathbf{x}; \mathbf{x}) = |\xi_1|^2 + |\xi_2|^2 + \cdots + |\xi_n|^2.$$

We have thus found a basis relative to which both quadratic forms $A(\mathbf{x}; \mathbf{x})$ and $B(\mathbf{x}; \mathbf{x})$ are expressible as sums of squares.

We now show how to find the numbers $\lambda_1, \lambda_2, \cdots, \lambda_n$ which appear in (2) above.

The matrices of the quadratic forms A and B have the following canonical form:

$$\mathscr{A} = \begin{bmatrix} \lambda_1 & 0 & \cdots & 0 \\ 0 & \lambda_2 & \cdots & 0 \\ \cdot & \cdot & \cdot & \cdot \\ 0 & 0 & \cdots & \lambda_n \end{bmatrix}, \qquad \mathscr{B} = \begin{bmatrix} 1 & 0 & \cdots & 0 \\ 0 & 1 & \cdots & 0 \\ \cdot & \cdot & \cdot & \cdot \\ 0 & 0 & \cdots & 1 \end{bmatrix}.$$

Consequently,

$$(4) \qquad \text{Det}\,(\mathscr{A} - \lambda\mathscr{B}) = (\lambda_1 - \lambda)(\lambda_2 - \lambda) \cdots (\lambda_n - \lambda).$$

Under a change of basis the matrices of the Hermitian quadratic forms A and B go over into the matrices $\mathscr{A}_1 = \mathscr{C}^*\mathscr{A}\mathscr{C}$ and $\mathscr{B}_1 = \mathscr{C}^*\mathscr{B}\mathscr{C}$. Hence, if $\mathbf{e}_1, \mathbf{e}_2, \cdots, \mathbf{e}_n$ is an arbitrary basis, then with respect to this basis

$$\text{Det}\,(\mathscr{A}_1 - \lambda\mathscr{B}_1) = \text{Det}\,\mathscr{C}^* \cdot \text{Det}\,(\mathscr{A} - \lambda\mathscr{B}) \cdot \text{Det}\,\mathscr{C},$$

i.e., Det $(\mathscr{A}_1 - \lambda\mathscr{B}_1)$ differs from (4) by a multiplicative constant. It follows that *the numbers* $\lambda_1, \lambda_2, \cdots, \lambda_n$ *are the roots of the equation*

$$\begin{vmatrix} a_{11} - \lambda b_{11} & a_{12} - \lambda b_{12} & \cdots & a_{1n} - \lambda b_{1n} \\ a_{21} - \lambda b_{21} & a_{22} - \lambda b_{22} & \cdots & a_{2n} - \lambda b_{2n} \\ \cdot & \cdot & \cdot & \cdot \\ a_{n1} - \lambda b_{n1} & a_{n2} - \lambda b_{n2} & \cdots & a_{nn} - \lambda b_{nn} \end{vmatrix} = 0,$$

[1] Orthonormal relative to the inner product $(\mathbf{x}, \mathbf{y}) = B(\mathbf{x}; \mathbf{y})$.

where $||a_{ik}||$ and $||b_{ik}||$ are the matrices of the quadratic forms $A(\mathbf{x}; \mathbf{x})$ and $B(\mathbf{x}; \mathbf{x})$ in some basis $\mathbf{e}_1, \mathbf{e}_2, \cdots, \mathbf{e}_n$.

NOTE: The following example illustrates that the requirement that one of the two forms be positive definite is essential. The two quadratic forms

$$A(\mathbf{x}; \mathbf{x}) = |\xi_1|^2 - |\xi_2|^2, \qquad B(\mathbf{x}; \mathbf{x}) = \xi_1\bar{\xi}_2 + \xi_2\bar{\xi}_1,$$

neither of which is positive definite, cannot be reduced simultaneously to a sum of squares. Indeed, the matrix of the first form is

$$\mathscr{A} = \begin{bmatrix} 1 & 0 \\ 0 & -1 \end{bmatrix}$$

and the matrix of the second form is

$$\mathscr{B} = \begin{bmatrix} 0 & 1 \\ 1 & 0 \end{bmatrix}.$$

Consider the matrix $\mathscr{A} - \lambda\mathscr{B}$, where λ is a real parameter. Its determinant is equal to $-(\lambda^2 + 1)$ and has no real roots. Therefore, in accordance with the preceding discussion, the two forms cannot be reduced simultaneously to a sum of squares.

§ *13. Unitary transformations*

In § 11 we defined a unitary transformation by the equation

$$\text{(1)} \qquad \mathrm{UU}^* = \mathrm{U}^*\mathrm{U} = \mathrm{E}.$$

This definition has a simple geometric interpretation, namely:

A unitary transformation U *on an n-dimensional Euclidean space* **R** *preserves inner products, i.e.,*

$$(\mathrm{U}\mathbf{x}, \mathrm{U}\mathbf{y}) = (\mathbf{x}, \mathbf{y})$$

for all $\mathbf{x}, \mathbf{y} \in \mathbf{R}$. *Conversely, any linear transformation* U *which preserves inner products is unitary (i.e., it satisfies condition* (1)).

Indeed, assume $\mathrm{U}^*\mathrm{U} = \mathrm{E}$. Then

$$(\mathrm{U}\mathbf{x}, \mathrm{U}\mathbf{y}) = (\mathbf{x}, \mathrm{U}^*\mathrm{U}\mathbf{y}) = (\mathbf{x}, \mathbf{y}).$$

Conversely, if for any vectors $\mathbf{x}$ and $\mathbf{y}$

$$(\mathrm{U}\mathbf{x}, \mathrm{U}\mathbf{y}) = (\mathbf{x}, \mathbf{y}),$$

then

$$(\mathrm{U}^*\mathrm{U}\mathbf{x}, \mathbf{y}) = (\mathbf{x}, \mathbf{y}),$$

that is

$$(\mathrm{U}^*\mathrm{U}\mathbf{x}, \mathbf{y}) = (\mathrm{E}\mathbf{x}, \mathbf{y}).$$

Since equality of bilinear forms implies equality of corresponding transformations, it follows that $U^*U = E$, i.e., U is unitary.

In particular, for $\mathbf{x} = \mathbf{y}$ we have

$$(U\mathbf{x}, U\mathbf{x}) = (\mathbf{x}, \mathbf{x}),$$

i.e., *a unitary transformation preserves the length of a vector.*

EXERCISE. Prove that a linear transformation which preserves length is unitary.

We shall now characterize the matrix of a unitary transformation. To do this, we select an orthonormal basis $\mathbf{e}_1, \mathbf{e}_2, \cdots, \mathbf{e}_n$. Let

$$\begin{bmatrix} a_{11} & a_{12} & \cdots & a_{1n} \\ a_{21} & a_{22} & \cdots & a_{2n} \\ \cdots & \cdots & \cdots & \cdots \\ a_{n1} & a_{n2} & \cdots & a_{nn} \end{bmatrix} \tag{2}$$

be the matrix of the transformation U relative to this basis. Then

$$\begin{bmatrix} \bar{a}_{11} & \bar{a}_{21} & \cdots & \bar{a}_{n1} \\ \bar{a}_{12} & \bar{a}_{22} & \cdots & \bar{a}_{n2} \\ \cdots & \cdots & \cdots & \cdots \\ \bar{a}_{1n} & \bar{a}_{2n} & \cdots & \bar{a}_{nn} \end{bmatrix} \tag{3}$$

is the matrix of the adjoint U^* of U.

The condition $UU^* = E$ implies that the product of the matrices (2) and (3) is equal to the unit matrix, that is,

$$\sum_{\alpha=1}^{n} a_{i\alpha}\bar{a}_{i\alpha} = 1, \qquad \sum_{\alpha=1}^{n} a_{i\alpha}\bar{a}_{k\alpha} = 0 \qquad (i \neq k). \tag{4}$$

Thus, relative to an orthonormal basis, the matrix of a unitary transformation U *has the following properties: the sum of the products of the elements of any row by the conjugates of the corresponding elements of any other row is equal to zero; the sum of the squares of the moduli of the elements of any row is equal to one.*

Making use of the condition $U^*U = E$ we obtain, in addition,

$$\sum_{\alpha=1}^{n} a_{\alpha i}\bar{a}_{\alpha i} = 1, \qquad \sum_{\alpha=1}^{n} a_{\alpha i}\bar{a}_{\alpha k} = 0 \qquad (i \neq k). \tag{5}$$

This condition is analogous to the preceding one, but refers to the columns rather than the rows of the matrix of U.

Condition (5) has a simple geometric meaning. Indeed, the inner product of the vectors

$$\mathrm{U}\mathbf{e}_i = a_{1i}\mathbf{e}_1 + a_{2i}\mathbf{e}_2 + \cdots + a_{ni}\mathbf{e}_n$$

and

$$\mathrm{U}\mathbf{e}_k = a_{1k}\mathbf{e}_1 + a_{2k}\mathbf{e}_2 + \cdots + a_{nk}\mathbf{e}_n$$

is equal to $\sum a_{\alpha i}\bar{a}_{\alpha k}$ (since we assumed $\mathbf{e}_1, \mathbf{e}_2, \cdots, \mathbf{e}_n$ to be an orthonormal basis). Hence

$$(6) \qquad (\mathrm{U}\mathbf{e}_i, \mathrm{U}\mathbf{e}_k) = \begin{cases} 1 & \text{for } i = k, \\ 0 & \text{for } i \neq k. \end{cases}$$

It follows that *a necessary and sufficient condition for a linear transformation* U *to be unitary is that it take an orthonormal basis* $\mathbf{e}_1, \mathbf{e}_2, \cdots, \mathbf{e}_n$ *into an orthonormal basis* $\mathrm{U}\mathbf{e}_1, \mathrm{U}\mathbf{e}_2, \cdots, \mathrm{U}\mathbf{e}_n$.

A matrix $||a_{ik}||$ whose elements satisfy condition (4) or, equivalently, condition (5) is called *unitary*. As we have shown unitary matrices are matrices of unitary transformations relative to an orthonormal basis. Since a transformation which takes an orthonormal basis into another orthonormal basis is unitary, the matrix of transition from an orthonormal basis to another orthonormal basis is also unitary.

We shall now try to find the simplest form of the matrix of a unitary transformation relative to some suitably chosen basis.

LEMMA 1. *The eigenvalues of a unitary transformation are in absolute value equal to one.*

Proof: Let $\mathbf{x}$ be an eigenvector of a unitary transformation U and let λ be the corresponding eigenvalue, i.e.,

$$\mathrm{U}\mathbf{x} = \lambda\mathbf{x}, \qquad \mathbf{x} \neq 0.$$

Then

$$(\mathbf{x}, \mathbf{x}) = (\mathrm{U}\mathbf{x}, \mathrm{U}\mathbf{x}) = (\lambda\mathbf{x}, \lambda\mathbf{x}) = \lambda\bar{\lambda}(\mathbf{x}, \mathbf{x}),$$

that is, $\lambda\bar{\lambda} = 1$ or $|\lambda| = 1$.

LEMMA 2. *Let* U *be a unitary transformation on an n-dimensional space* $\mathbf{R}$ *and* $\mathbf{e}$ *its eigenvector, i.e.,*

$$\mathrm{U}\mathbf{e} = \lambda\mathbf{e}, \quad \mathbf{e} \neq \mathbf{0}.$$

Then the $(n-1)$*-dimensional subspace* $\mathbf{R}_1$ *of* $\mathbf{R}$ *consisting of all vectors* $\mathbf{x}$ *orthogonal to* $\mathbf{e}$ *is invariant under* U.

Proof: Let $\mathbf{x} \in \mathbf{R}_1$, i.e., $(\mathbf{x}, \mathbf{e}) = 0$. We shall show that $\mathrm{U}\mathbf{x} \in \mathbf{R}_1$, i.e., $(\mathrm{U}\mathbf{x}, \mathbf{e}) = 0$. Indeed,

$$(\mathrm{U}\mathbf{x}, \mathrm{U}\mathbf{e}) = (\mathrm{U}^*\mathrm{U}\mathbf{x}, \mathbf{e}) = (\mathbf{x}, \mathbf{e}) = 0.$$

Since $\mathrm{U}\mathbf{e} = \lambda\mathbf{e}$, it follows that $\bar{\lambda}(\mathrm{U}\mathbf{x}, \mathbf{e}) = 0$. By Lemma 1, $\lambda \neq 0$, hence $(\mathrm{U}\mathbf{x}, \mathbf{e}) = 0$, i.e., $\mathrm{U}\mathbf{x} \in \mathbf{R}_1$. Thus, the subspace $\mathbf{R}_1$ is indeed invariant under U.

THEOREM 1. *Let* U *be a unitary transformation defined on an n-dimensional Euclidean space* **R**. *Then* U *has n pairwise orthogonal eigenvectors. The corresponding eigenvalues are in absolute value equal to one.*

Proof: In view of Theorem 1, § 10, the transformation U as a linear transformation has at least one eigenvector. Denote this vector by $\mathbf{e}_1$. By Lemma 2, the $(n-1)$-dimensional subspace $\mathbf{R}_1$ of all vectors of **R** which are orthogonal to $\mathbf{e}_1$ is invariant under U. Hence $\mathbf{R}_1$ contains at least one eigenvector $\mathbf{e}_2$ of U. Denote by $\mathbf{R}_2$ the invariant subspace consisting of all vectors of $\mathbf{R}_1$ orthogonal to $\mathbf{e}_2$. $\mathbf{R}_2$ contains at least one eigenvector $\mathbf{e}_3$ of U, etc. Proceeding in this manner we obtain n pairwise orthogonal eigenvectors $\mathbf{e}_1, \mathbf{e}_2, \cdots, \mathbf{e}_n$ of the transformation U. By Lemma 1 the eigenvalues corresponding to these eigenvectors are in absolute value equal to one.

THEOREM 2. *Let* U *be a unitary transformation on an n-dimensional Euclidean space* **R**. *Then there exists an orthonormal basis in* **R** *relative to which the matrix of the transformation* U *is diagonal, i.e., has the form*

$$\begin{bmatrix} \lambda_1 & 0 & \cdots & 0 \\ 0 & \lambda_2 & \cdots & 0 \\ 0 & 0 & \cdots & \lambda_n \end{bmatrix}. \tag{7}$$

The numbers $\lambda_1, \lambda_2, \cdots, \lambda_n$ *are in absolute value equal to one.*

Proof: Let U be a unitary transformation. We claim that the n pairwise orthogonal eigenvectors constructed in the preceding theorem constitute the desired basis. Indeed,

$$\begin{aligned} \mathrm{U}\mathbf{e}_1 &= \lambda_1\mathbf{e}_1, \\ \mathrm{U}\mathbf{e}_2 &= \lambda_2\mathbf{e}_2, \\ &\cdots\cdots \\ \mathrm{U}\mathbf{e}_n &= \lambda_n\mathbf{e}_n, \end{aligned}$$

and, therefore, the matrix of U relative to the basis $\mathbf{e}_1, \mathbf{e}_2, \cdots, \mathbf{e}_n$ has form (7). By Lemma 1 the numbers $\lambda_1, \lambda_2, \cdots, \lambda_n$ are in absolute value equal to one. This proves the theorem.

EXERCISES. *1.* Prove the converse of Theorem 2, i.e., if the matrix of U has form (7) relative to some orthogonal basis then U is unitary.

2. Prove that if A is a self-adjoint transformation then the transformation $(\mathrm{A} - i\mathrm{E})^{-1} \cdot (\mathrm{A} + i\mathrm{E})$ exists and is unitary.

Since the matrix of transition from one orthonormal basis to another is unitary we can give the following matrix interpretation to the result obtained in this section.

Let $\mathscr{U}$ be a unitary matrix. Then there exists a unitary matrix $\mathscr{V}$ such that

$$\mathscr{U} = \mathscr{V}^{-1}\mathscr{D}\mathscr{V},$$

where $\mathscr{D}$ is a diagonal matrix whose non-zero elements are equal in absolute value to one.

Analogously, the main result of para. 1, § 12, can be given the following matrix interpretation.

Let $\mathscr{A}$ be a Hermitian matrix. Then $\mathscr{A}$ can be represented in the form

$$\mathscr{A} = \mathscr{V}^{-1}\mathscr{D}\mathscr{V},$$

where $\mathscr{V}$ is a unitary matrix and $\mathscr{D}$ a diagonal matrix whose non-zero elements are real.

§ *14.* *Commutative linear transformations. Normal transformations*

1. *Commutative transformations.* We have shown (§ 12) that for each self-adjoint transformation there exists an orthonormal basis relative to which the matrix of the transformation is diagonal. It may turn out that given a number of self-adjoint transformations, we can find a basis relative to which all these transformations are represented by diagonal matrices. We shall now discuss conditions for the existence of such a basis. We first consider the case of two transformations.

LEMMA 1. *Let* A *and* B *be two commutative linear transformations, i.e., let*

$$\mathrm{AB} = \mathrm{BA}.$$

Then the eigenvectors of A *which correspond to a given eigenvalue* λ *of* A *form (together with the null vector) a subspace* $\mathbf{R}_\lambda$ *invariant under the transformation* B.

Proof: We have to show that if

$$\mathbf{x} \in \mathbf{R}_\lambda, \text{ i.e., } \mathrm{A}\mathbf{x} = \lambda \mathbf{x},$$

then

$$\mathrm{B}\mathbf{x} \in \mathbf{R}_\lambda, \quad \text{i.e., } \mathrm{AB}\mathbf{x} = \lambda \mathrm{B}\mathbf{x}.$$

Since AB = BA, we have

$$\mathrm{AB}\mathbf{x} = \mathrm{BA}\mathbf{x} = \mathrm{B}\lambda\mathbf{x} = \lambda\mathrm{B}\mathbf{x},$$

which proves our lemma.

LEMMA 2. *Any two commutative transformations have a common eigenvector.*

Proof: Let AB = BA and let $\mathbf{R}_\lambda$ be the subspace consisting of all vectors $\mathbf{x}$ for which $\mathrm{A}\mathbf{x} = \lambda\mathbf{x}$, where λ is an eigenvalue of A. By Lemma 1, $\mathbf{R}_\lambda$ is invariant under B. Hence $\mathbf{R}_\lambda$ contains a vector $\mathbf{x}_0$ which is an eigenvector of B. $\mathbf{x}_0$ is also an eigenvector of A, since by assumption all the vectors of $\mathbf{R}_\lambda$ are eigenvectors of A.

NOTE: If AB = BA we cannot claim that every eigenvector of A is also an eigenvector of B. For instance, if A is the identity transformation E, B a linear transformation other than E and $\mathbf{x}$ a vector which is not an eigenvector of B, then $\mathbf{x}$ is an eigenvector of E, EB = BE and $\mathbf{x}$ is not an eigenvector of B.

THEOREM 1. *Let* A *and* B *be two linear self-adjoint transformations defined on a complex n-dimensional vector space* $\mathbf{R}$. *A necessary and sufficient condition for the existence of an orthogonal basis in* $\mathbf{R}$ *relative to which the transformations* A *and* B *are represented by diagonal matrices is that* A *and* B *commute.*

Sufficiency: Let AB = BA. Then, by Lemma 2, there exists a vector $\mathbf{e}_1$ which is an eigenvector of both A and B, i.e.,

$$\mathrm{A}\mathbf{e}_1 = \lambda_1 \mathbf{e}_1, \qquad \mathrm{B}\mathbf{e}_1 = \mu_1 \mathbf{e}_1.$$

The $(n - 1)$-dimensional subspace $\mathbf{R}_1$ orthogonal to $\mathbf{e}_1$ is invariant under A and B (cf. Lemma 2, § 12). Now consider A and B on $\mathbf{R}_1$ only. By Lemma 2, there exists a vector $\mathbf{e}_2$ in $\mathbf{R}_1$ which is an eigenvector of A and B:

$$\mathrm{A}\mathbf{e}_2 = \lambda_2 \mathbf{e}_2, \qquad \mathrm{B}\mathbf{e}_2 = \mu_2 \mathbf{e}_2.$$

All vectors of $\mathbf{R}_1$ which are orthogonal to $\mathbf{e}_2$ form an $(n-2)$-dimensional subspace invariant under A and B, etc. Proceeding in this way we get n pairwise orthogonal eigenvectors $\mathbf{e}_1, \mathbf{e}_2, \cdots, \mathbf{e}_n$ of A and B:

$$\mathrm{A}\mathbf{e}_i = \lambda_i \mathbf{e}_i, \qquad \mathrm{B}\mathbf{e}_i = \mu_i \mathbf{e}_i \qquad (i = 1, \cdots, n).$$

Relative to $\mathbf{e}_1, \mathbf{e}_2, \cdots, \mathbf{e}_n$ the matrices of A and B are diagonal. This completes the sufficiency part of the proof.

Necessity: Assume that the matrices of A and B are diagonal relative to some orthogonal basis. It follows that these matrices commute. But then the transformations themselves commute.

EXERCISE. Let U_1 and U_2 be two commutative unitary transformations. Prove that there exists a basis relative to which the matrices of U_1 and U_2 are diagonal.

NOTE: Theorem 1 can be generalized to any set of pairwise commutative self-adjoint transformations. The proof follows that of Theorem 1 but instead of Lemma 2 the following Lemma is made use of:

LEMMA 2′. *The elements of any set of pairwise commutative transformations on a vector space* **R** *have a common eigenvector.*

Proof: The proof is by induction on the dimension of the space **R**. In the case of one-dimensional space $(n = 1)$ the lemma is obvious. We assume that it is true for spaces of dimension $< n$ and prove it for an n-dimensional space.

If every vector of **R** is an eigenvector of all the transformations A, B, C, $\cdots$ in our set [8] our lemma is proved. Assume therefore that there exists a vector in **R** which is not an eigenvector of the transformation A, say.

Let $\mathbf{R}_1$ be the set of all eigenvectors of A corresponding to some eigenvalue λ of A. By Lemma 1, $\mathbf{R}_1$ is invariant under each of the transformations B, C, $\cdots$ (obviously, $\mathbf{R}_1$ is also invariant under A). Furthermore, $\mathbf{R}_1$ is a subspace different from the null space and the whole space. Hence $\mathbf{R}_1$ is of dimension $\leqq n - 1$. Since, by assumption, our lemma is true for spaces of dimension $< n$, $\mathbf{R}_1$ must contain a vector which is an eigenvector of the transformations A, B, C, $\cdots$. This proves our lemma.

2. *Normal transformations.* In §§ 12 and 13 we considered two classes of linear transformations which are represented in a suitable orthonormal basis by a diagonal matrix. We shall now characterize all transformations with this property.

THEOREM 2. *A necessary and sufficient condition for the existence*

[8] This means that the transformations A, B, C, $\cdots$ are multiples of the identity transformation.

of an orthogonal basis relative to which a transformation A *is represented by a diagonal matrix is*

$$AA^* = A^*A$$

(*such transformations are said to be normal, cf.* § 11).

Necessity: Let the matrix of the transformation A be diagonal relative to some orthonormal basis, i.e., let the matrix be of the form

$$\begin{bmatrix} \lambda_1 & 0 & \cdots & 0 \\ 0 & \lambda_2 & \cdots & 0 \\ \cdots & \cdots & \cdots & \cdots \\ 0 & 0 & \cdots & \lambda_n \end{bmatrix}.$$

Relative to such a basis the matrix of the transformation A* has the form

$$\begin{bmatrix} \bar{\lambda}_1 & 0 & \cdots & 0 \\ 0 & \bar{\lambda}_2 & \cdots & 0 \\ \cdots & \cdots & \cdots & \cdots \\ 0 & 0 & \cdots & \bar{\lambda}_n \end{bmatrix}.$$

Since the matrices of A and A* are diagonal they commute. It follows that A and A* commute.

Sufficiency: Assume that A and A* commute. Then by Lemma 2 there exists a vector $\mathbf{e}_1$ which is an eigenvector of A and A*, i.e.,

$$A\mathbf{e}_1 = \lambda_1 \mathbf{e}_1, \qquad A^*\mathbf{e}_1 = \mu_1 \mathbf{e}_1. \text{ [9]}$$

The $(n - 1)$-dimensional subspace $\mathbf{R}_1$ of vectors orthogonal to $\mathbf{e}_1$ is invariant under A as well as under A*. Indeed, let $\mathbf{x} \in \mathbf{R}_1$, i.e., $(\mathbf{x}, \mathbf{e}_1) = 0$. Then

$$(A\mathbf{x}, \mathbf{e}_1) = (\mathbf{x}, A^*\mathbf{e}_1) = (\mathbf{x}, \mu_1\mathbf{e}_1) = \bar{\mu}_1(\mathbf{x}, \mathbf{e}_1) = 0,$$

that is, $A\mathbf{x} \in \mathbf{R}_1$. This proves that $\mathbf{R}_1$ is invariant under A. The invariance of $\mathbf{R}_1$ under A* is proved in an analogous manner.

Applying now Lemma 2 to $\mathbf{R}_1$, we can claim that $\mathbf{R}_1$ contains a vector $\mathbf{e}_2$ which is an eigenvector of A and A*. Let $\mathbf{R}_2$ be the $(n - 2)$-dimensional subspace of vectors from $\mathbf{R}_2$ orthogonal to $\mathbf{e}_2$, etc. Continuing in this manner we construct n pairwise orthogonal vectors $\mathbf{e}_1, \mathbf{e}_2, \cdots, \mathbf{e}_n$ which are eigenvectors of A and A*.

[9] EXERCISE. Prove that $\mu_1 = \bar{\lambda}_1$.

The vectors $\mathbf{e}_1, \mathbf{e}_2, \cdots, \mathbf{e}_n$ form an orthogonal basis relative to which both A and A* are represented by diagonal matrices.

An alternative sufficiency proof. Let

$$\mathrm{A}_1 = \frac{\mathrm{A} + \mathrm{A}^*}{2}, \quad \mathrm{A}_2 = \frac{\mathrm{A} - \mathrm{A}^*}{2i}.$$

The transformations A_1 and A_2 are self-adjoint. If A and A* commute then so do A_1 and A_2. By Theorem 1, there exists an orthonormal basis in which A_1 and A_2 are represented by diagonal matrices. But then the same is true of $\mathrm{A} = \mathrm{A}_1 + i\mathrm{A}_2$.

Note that if A is a self-adjoint transformation then

$$\mathrm{AA}^* = \mathrm{A}^*\mathrm{A} = \mathrm{A}^2,$$

i.e., A is normal. A unitary transformation U is also normal since $\mathrm{UU}^* = \mathrm{U}^*\mathrm{U} = \mathrm{E}$. Thus some of the results obtained in para. 1, § 12 and § 13 are special cases of Theorem 2.

EXERCISES. *1.* Prove that the matrices of a set of normal transformations any two of which commute are simultaneously diagonable.

2. Prove that a normal transformation A can be written in the form

$$\mathrm{A} = \mathrm{HU} = \mathrm{UH},$$

where H is self-adjoint, U unitary and where H and U commute.

Hint: Select a basis relative to which A and A* are diagonable.

3. Prove that if $\mathrm{A} = \mathrm{HU}$, where H and U commute, H is self-adjoint and U unitary, then A is normal.

§ *15.* Decomposition of a linear transformation into a product of a unitary and self-adjoint transformation

Every complex number can be written as a product of a positive number and a number whose absolute value is one (the so-called trigonometric form of a complex number). We shall now derive an analogous result for linear transformations.

Unitary transformations are the analog of numbers of absolute value one. The analog of positive numbers are the so-called positive definite linear transformations.

DEFINITION 1. *A linear transformation* H *is called positive definite if it is self-adjoint and if* $(\mathrm{H}\mathbf{x}, \mathbf{x}) \geqq 0$ *for all* $\mathbf{x}$.

THEOREM 1. *Every non-singular linear transformation* A *can be represented in the form*

$$A = HU \qquad (\text{or } A = U_1 H_1),$$

where $H(H_1)$ *is a non-singular positive definite transformation and* $U(U_1)$ *a unitary transformation.*

We shall first assume the theorem true and show how to find the necessary H and U. This will suggest a way of proving the theorem.

Thus, let $A = HU$, where U is unitary and H is a non-singular positive definite transformation. H is easily expressible in terms of A. Indeed,

$$A^* = U^* H^* = U^{-1} H,$$

so that

$$AA^* = H^2.$$

Consequently, in order to find H one has to "extract the square root" of AA^*. Having found H, we put $U = H^{-1}A$.

Before proving Theorem 1 we establish three lemmas.

LEMMA 1. *Given any linear transformation* A, *the transformation* AA^* *is positive definite.* If A *is non-singular then so is* AA^*.

Proof: The transformation AA^* is positive definite. Indeed,

$$(AA^*)^* = A^{**}A^* = AA^*,$$

that is, AA^* is self-adjoint. Furthermore,

$$(AA^*\mathbf{x}, \mathbf{x}) = (A^*\mathbf{x}, A^*\mathbf{x}) \geqq 0,$$

for all $\mathbf{x}$. Thus AA^* is positive definite.

If A is non-singular, then the determinant of the matrix $||a_{ik}||$ of the transformation A relative to any orthogonal basis is different from zero. The determinant of the matrix $||\bar{a}_{ki}||$ of the transformation A^* relative to the same basis is the complex conjugate of the determinant of the matrix $||a_{ik}||$. Hence the determinant of the matrix of AA^* is different from zero, which means that AA^* is non-singular.

LEMMA 2. *The eigenvalues of a positive definite transformation* B *are non-negative. Conversely, if all the eigenvalues of a self-adjoint transformation* B *are non-negative then* B *is positive definite.*

Proof. Let B be positive definite and let $\mathrm{B}\mathbf{e} = \lambda\mathbf{e}$. Then

$$(\mathrm{B}\mathbf{e}, \mathbf{e}) = \lambda(\mathbf{e}, \mathbf{e}).$$

Since $(\mathrm{B}\mathbf{e}, \mathbf{e}) \geqq 0$ and $(\mathbf{e}, \mathbf{e}) > 0$, it follows that $\lambda \geqq 0$.

Conversely, assume that all the eigenvalues of a self-adjoint transformation B are non-negative. Let $\mathbf{e}_1, \mathbf{e}_2, \cdots, \mathbf{e}_n$ be an orthonormal basis consisting of the eigenvectors of B. Let

$$\mathbf{x} = \xi_1\mathbf{e}_1 + \xi_2\mathbf{e}_2 + \cdots + \xi_n\mathbf{e}_n,$$

be any vector of **R**. Then

$$(1)\quad \begin{aligned} &(\mathrm{B}\mathbf{x}, \mathbf{x}) \\ &= (\xi_1\mathrm{B}\mathbf{e}_1 + \xi_2\mathrm{B}\mathbf{e}_2 + \cdots + \xi_n\mathrm{B}\mathbf{e}_n, \xi_1\mathbf{e}_1 + \xi_2\mathbf{e}_2 + \cdots + \xi_n\mathbf{e}_n) \\ &= (\xi_1\lambda_1\mathbf{e}_1 + \xi_2\lambda_2\mathbf{e}_2 + \cdots + \xi_n\lambda_n\mathbf{e}_n, \xi_1\mathbf{e}_1 + \xi_2\mathbf{e}_2 + \cdots + \xi_n\mathbf{e}_n) \\ &= \lambda_1|\xi_1|^2 + \lambda_2|\xi_2|^2 + \cdots + \lambda_n|\xi_n|^2. \end{aligned}$$

Since all the λ_i are non-negative it follows that $(\mathrm{B}\mathbf{x}, \mathbf{x}) \geqq 0$.

NOTE: It is clear from equality (1) that if all the λ_i are positive then the transformation B is non-singular and, conversely, if B is positive definite and non-singular then the λ_i are positive.

LEMMA 3. *Given any positive definite transformation* B, *there exists a positive definite transformation* H *such that* $\mathrm{H}^2 = \mathrm{B}$ (*in this case we write* $\mathrm{H} = \sqrt{\mathrm{B}} = \mathrm{B}^{\frac{1}{2}}$). *In addition, if* B *is non-singular then* H *is non-singular.*

Proof: We select in **R** an orthogonal basis relative to which B is of the form

$$\mathrm{B} = \begin{bmatrix} \lambda_1 & 0 & \cdots & 0 \\ 0 & \lambda_2 & \cdots & 0 \\ \cdots & \cdots & \cdots & \cdots \\ 0 & 0 & \cdots & \lambda_n \end{bmatrix},$$

where $\lambda_1, \lambda_2, \cdots, \lambda_n$ are the eigenvalues of B. By Lemma 2 all $\lambda_i \geqq 0$. Put

$$\mathrm{H} = \begin{bmatrix} \sqrt{\lambda_1} & 0 & \cdots & 0 \\ 0 & \sqrt{\lambda_2} & \cdots & 0 \\ \cdots & \cdots & \cdots & \cdots \\ 0 & 0 & \cdots & \sqrt{\lambda_n} \end{bmatrix}.$$

Applying Lemma 2 again we conclude that H is positive definite.

Furthermore, if B is non-singular, then (cf. note to Lemma 2) $\lambda_i > 0$. Hence $\sqrt{\lambda_i} > 0$ and H is non-singular.

We now prove Theorem 1. Let A be a non-singular linear transformation. Let

$$H = \sqrt{(AA^*)}.$$

In view of Lemmas 1 and 3, H is a non-singular positive definite transformation. If

$$U = H^{-1}A, \tag{2}$$

then U is unitary. Indeed.

$$UU^* = H^{-1}A(H^{-1}A)^* = H^{-1}AA^*H^{-1} = H^{-1}H^2H^{-1} = E.$$

Making use of eq. (2) we get $A = HU$. This completes the proof of Theorem 1.

The operation of extracting the square root of a transformation can be used to prove the following theorem:

THEOREM. *Let* A *be a non-singular positive definite transformation and let* B *be a self-adjoint transformation. Then the eigenvalues of the transformation* AB *are real.*

Proof: We know that the transformations

$$X = AB \quad \text{and} \quad C^{-1}XC$$

have the same characteristic polynomials and therefore the same eigenvalues. If we can choose C so that $C^{-1}XC$ is self-adjoint, then $C^{-1}XC$ and $X = AB$ will both have real eigenvalues. A suitable choice for C is $C = A^{\frac{1}{2}}$. Then

$$C^{-1}XC = A^{-\frac{1}{2}}ABA^{\frac{1}{2}} = A^{\frac{1}{2}}BA^{\frac{1}{2}},$$

which is easily seen to be self-adjoint. Indeed,

$$(A^{\frac{1}{2}}BA^{\frac{1}{2}})^* = (A^{\frac{1}{2}})^*B^*(A^{\frac{1}{2}})^* = A^{\frac{1}{2}}BA^{\frac{1}{2}}.$$

This completes the proof.

EXERCISE. Prove that if A and B are positive definite transformations, at least one of which is non-singular, then the transformation AB has non-negative eigenvalues.

§ 16. *Linear transformations on a real Euclidean space*

This section will be devoted to a discussion of linear transformations defined on a real space. For the purpose of this discussion the reader need only be familiar with the material of §§ 9 through 11 of this chapter.

1. The concepts of invariant subspace, eigenvector, and eigenvalue introduced in § 10 were defined for a vector space over an arbitrary field and are therefore relevant in the case of a real vector space. In § 10 we proved that in a complex vector space every linear transformation has at least one eigenvector (one-dimensional invariant subspace). This result which played a fundamental role in the development of the theory of complex vector spaces does not apply in the case of real spaces. Thus, a rotation of the plane about the origin by an angle different from $k\pi$ is a linear transformation which does not have any one-dimensional invariant subspace. However, we can state the following

THEOREM 1. *Every linear transformation in a real vector space* **R** *has a one-dimensional or two-dimensional invariant subspace.*

Proof: Let $\mathbf{e}_1, \mathbf{e}_2, \cdots, \mathbf{e}_n$ be a basis in **R** and let $||a_{ik}||$ be the matrix of A relative to this basis.

Consider the system of equations

$$(1)\qquad \begin{cases} a_{11}\xi_1 + a_{12}\xi_2 + \cdots + a_{1n}\xi_n = \lambda\xi_1, \\ a_{21}\xi_1 + a_{22}\xi_2 + \cdots + a_{2n}\xi_n = \lambda\xi_2, \\ \cdots\cdots\cdots\cdots\cdots\cdots\cdots\cdots \\ a_{n1}\xi_1 + a_{n2}\xi_2 + \cdots + a_{nn}\xi_n = \lambda\xi_n. \end{cases}$$

The system (1) has a non-trivial solution if and only if

$$\begin{vmatrix} a_{11} - \lambda & a_{12} & \cdots & a_{1n} \\ a_{21} & a_{22} - \lambda & \cdots & a_{2n} \\ \cdots & \cdots & \cdots & \cdots \\ a_{n1} & a_{n2} & \cdots & a_{nn} - \lambda \end{vmatrix} = 0.$$

This equation is an nth order polynomial equation in λ with real coefficients. Let λ_0 be one of its roots. There arise two possibilities:

a. λ_0 is a real root. Then we can find numbers $\xi_1^0, \xi_2^0, \cdots, \xi_n^0$ not all zero which are a solution of (1). These numbers are the coordinates of some vector **x** relative to the basis $\mathbf{e}_1, \mathbf{e}_2, \cdots, \mathbf{e}_n$. We can thus rewrite (1) in the form

$$A\mathbf{x} = \lambda_0\mathbf{x},$$

i.e., the vector **x** spans a one-dimensional invariant subspace.

b. $\lambda_0 = \alpha + i\beta, \beta \neq 0$. Let

$$\xi_1 + i\eta_1, \xi_2 + i\eta_2, \cdots, \xi_n + i\eta_n$$

be a solution of (1). Replacing $\xi_1, \xi_2, \cdots, \xi_n$ in (1) by these numbers and separating the real and imaginary parts we get

$$(2) \qquad \begin{cases} a_{11}\xi_1 + a_{12}\xi_2 + \cdots + a_{1n}\xi_n = \alpha\xi_1 - \beta\eta_1, \\ a_{21}\xi_1 + a_{22}\xi_2 + \cdots + a_{2n}\xi_n = \alpha\xi_2 - \beta\eta_2, \\ \cdots\cdots\cdots\cdots\cdots\cdots\cdots\cdots\cdots\cdots \\ a_{n1}\xi_1 + a_{n2}\xi_2 + \cdots + a_{nn}\xi_n = a\xi_n - \beta\eta_n, \end{cases}$$

and

$$(2)' \qquad \begin{cases} a_{11}\eta_1 + a_{12}\eta_2 + \cdots + a_{1n}\eta_n = \alpha\eta_1 + \beta\xi_1, \\ a_{21}\eta_1 + a_{22}\eta_2 + \cdots + a_{2n}\eta_n = \alpha\eta_2 + \beta\xi_2, \\ \cdots\cdots\cdots\cdots\cdots\cdots\cdots\cdots\cdots\cdots \\ a_{n1}\eta_1 + a_{2n}\eta_2 + \cdots + a_{nn}\eta_n = \alpha\eta_n + \beta\xi_n. \end{cases}$$

The numbers $\xi_1, \xi_2, \cdots, \xi_n$ $(\eta_1, \eta_2, \cdots, \eta_n)$ are the coordinates of some vector $\mathbf{x}$ $(\mathbf{y})$ in $\mathbf{R}$. Thus the relations (2) and (2′) can be rewritten as follows

$$(3) \qquad \mathrm{A}\mathbf{x} = \alpha\mathbf{x} - \beta\mathbf{y}; \qquad \mathrm{A}\mathbf{y} = \alpha\mathbf{y} + \beta\mathbf{x}.$$

Equations (3) imply that the two dimensional subspace spanned by the vectors $\mathbf{x}$ and $\mathbf{y}$ is invariant under A.

In the sequel we shall make use of the fact that in a two-dimensional invariant subspace associated with the root $\lambda = \alpha + i\beta$ the transformation has form (3).

EXERCISE. Show that in an odd-dimensional space (in particular, three-dimensional) every transformation has a one-dimensional invariant subspace.

2. *Self-adjoint transformations*

DEFINITION 1. *A linear transformation* A *defined on a real Euclidean space* $\mathbf{R}$ *is said to be self-adjoint if*

$$(4) \qquad (\mathrm{A}\mathbf{x}, \mathbf{y}) = (\mathbf{x}, \mathrm{A}\mathbf{y})$$

for any vectors $\mathbf{x}$ *and* $\mathbf{y}$.

Let $\mathbf{e}_1, \mathbf{e}_2, \cdots, \mathbf{e}_n$ be an orthonormal basis in $\mathbf{R}$ and let

$\mathbf{x} = \xi_1\mathbf{e}_1 + \xi_2\mathbf{e}_2 + \cdots + \xi_n\mathbf{e}_n, \quad \mathbf{y} = \eta_1\mathbf{e}_1 + \eta_2\mathbf{e}_2 + \cdots + \eta_n\mathbf{e}_n.$

Furthermore, let ζ_i be the coordinates of the vector $\mathbf{z} = \mathrm{A}\mathbf{x}$, i.e.,

$$\zeta_i = \sum_{k=1}^{n} a_{ik}\xi_k,$$

where $||a_{ik}||$ is the matrix of A relative to the basis $\mathbf{e}_1, \mathbf{e}_2, \cdots, \mathbf{e}_n$. It follows that

$$(\mathrm{A}\mathbf{x}, \mathbf{y}) = (\mathbf{z}, \mathbf{y}) = \sum_{i=1}^{n} \zeta_i \eta_i = \sum_{i,k=1}^{n} a_{ik}\xi_k\eta_i.$$

Similarly,

$$(\mathbf{x}, \mathrm{A}\mathbf{y}) = \sum_{i,k=1}^{n} a_{ik}\xi_i\eta_k. \tag{5}$$

Thus, condition (4) is equivalent to

$$a_{ik} = a_{ki}.$$

To sum up, *for a linear transformation to be self-adjoint it is necessary and sufficient that its matrix relative to an orthonormal basis be symmetric.*

Relative to an arbitrary basis every symmetric bilinear form $A(\mathbf{x}; \mathbf{y})$ is represented by

$$A(\mathbf{x}; \mathbf{y}) = \sum_{i,k=1}^{n} a_{ik}\xi_i\eta_k \tag{6}$$

where $a_{ik} = a_{ki}$. Comparing (5) and (6) we obtain the following result:

Given a symmetric bilinear form $A(\mathbf{x}; \mathbf{y})$ *there exists a self-adjoint transformation* A *such that*

$$A(\mathbf{x}; \mathbf{y}) = (\mathrm{A}\mathbf{x}, \mathbf{y}).$$

We shall make use of this result in the proof of Theorem 3 of this section.

We shall now show that given a self-adjoint transformation there exists an orthogonal basis relative to which the matrix of the transformation is diagonal. The proof of this statement will be based on the material of para. 1. A different proof which does not depend on the results of para. 1 and is thus independent of the theorem asserting the existence of the root of an algebraic equation is given in § 17.

We first prove two lemmas.

LEMMA 1. *Every self-adjoint transformation has a one-dimensional invariant subspace.*

Proof: According to Theorem 1 of this section, to every real root λ of the characteristic equation there corresponds a one-dimensional invariant subspace and to every complex root λ, a two-dimensional invariant subspace. Thus, to prove Lemma 1 we need only show that all the roots of a self-adjoint transformation are real.

Suppose that $\lambda = \alpha + i\beta$, $\beta \neq 0$. In the proof of Theorem 1 we constructed two vectors $\mathbf{x}$ and $\mathbf{y}$ such that

$$\begin{aligned} A\mathbf{x} &= \alpha\mathbf{x} - \beta\mathbf{y}, \\ A\mathbf{y} &= \beta\mathbf{x} + \alpha\mathbf{y}. \end{aligned}$$

But then

$$\begin{aligned} (A\mathbf{x}, \mathbf{y}) &= \alpha(\mathbf{x}, \mathbf{y}) - \beta(\mathbf{y}, \mathbf{y}) \\ (\mathbf{x}, A\mathbf{y}) &= \beta(\mathbf{x}, \mathbf{x}) + \alpha(\mathbf{x}, \mathbf{y}). \end{aligned}$$

Subtracting the first equation from the second we get [note that $(A\mathbf{x}, \mathbf{y}) = (\mathbf{x}, A\mathbf{y})$]

$$0 = 2\beta[(\mathbf{x}, \mathbf{x}) + (\mathbf{y}, \mathbf{y})].$$

Since $(\mathbf{x}, \mathbf{x}) + (\mathbf{y}, \mathbf{y}) \neq 0$, it follows that $\beta = 0$. Contradiction.

LEMMA 2. *Let* A *be a self-adjoint transformation and* $\mathbf{e}_1$ *an eigenvector of* A. *Then the totality* $\mathbf{R}'$ *of vectors orthogonal to* $\mathbf{e}_1$ *forms an* $(n - 1)$*-dimensional invariant subspace.*

Proof: It is clear that the totality $\mathbf{R}'$ of vectors $\mathbf{x}$, $\mathbf{x} \in \mathbf{R}$, orthogonal to $\mathbf{e}_1$ forms an $(n - 1)$-dimensional subspace. We show that $\mathbf{R}'$ is invariant under A.

Thus, let $\mathbf{x} \in \mathbf{R}'$, i.e., $(\mathbf{x}, \mathbf{e}_1) = 0$. Then

$$(A\mathbf{x}, \mathbf{e}_1) = (\mathbf{x}, A\mathbf{e}_1) = (\mathbf{x}, \lambda\mathbf{e}_1) = \lambda(\mathbf{x}, \mathbf{e}_1) = 0,$$

i.e., $A\mathbf{x} \in \mathbf{R}'$.

THEOREM 2. *There exists an orthonormal basis relative to which the matrix of a self-adjoint transformation* A *is diagonal.*

Proof: By Lemma 1, the transformation A has at least one eigenvector $\mathbf{e}_1$.

Denote by $\mathbf{R}'$ the subspace consisting of vectors orthogonal to $\mathbf{e}_1$. Since $\mathbf{R}'$ is invariant under A, it contains (again, by Lemma 1) an eigenvector $\mathbf{e}_2$ of A, etc. In this manner we obtain n pairwise orthogonal eigenvectors $\mathbf{e}_1, \mathbf{e}_2, \cdots, \mathbf{e}_n$.

Since

$$\mathrm{A}\mathbf{e}_i = \lambda_i \mathbf{e}_i \qquad (i = 1, 2, \cdots, n),$$

the matrix of A relative to the $\mathbf{e}_i$ is of the form

$$\begin{bmatrix} \lambda_1 & 0 & \cdots & 0 \\ 0 & \lambda_2 & \cdots & 0 \\ \cdots & \cdots & \cdots & \cdots \\ 0 & 0 & \cdots & \lambda_n \end{bmatrix}.$$

3. *Reduction of a quadratic form to a sum of squares relative to an orthogonal basis (reduction to principal axes).* Let $A(\mathbf{x}; \mathbf{y})$ be a symmetric bilinear form on an n-dimensional Euclidean space. We showed earlier that to each symmetric bilinear form $A(\mathbf{x}; \mathbf{y})$ there corresponds a linear self-adjoint transformation A such that $A(\mathbf{x}; \mathbf{y}) = (\mathrm{A}\mathbf{x}, \mathbf{y})$. According to Theorem 2 of this section there exists an orthonormal basis $\mathbf{e}_1, \mathbf{e}_2, \cdots, \mathbf{e}_n$ consisting of the eigenvectors of the transformation A (i.e., of vectors such that $\mathrm{A}\mathbf{e}_i = \lambda \mathbf{e}_i$). With respect to such a basis

$$\begin{aligned} &\mathrm{A}(\mathbf{x}; \mathbf{y}) = (\mathrm{A}\mathbf{x}, \mathbf{y}) \\ &= (\mathrm{A}(\xi_1\mathbf{e}_1 + \xi_2\mathbf{e}_2 + \cdots + \xi_n\mathbf{e}_n), \eta_1\mathbf{e}_1 + \eta_2\mathbf{e}_2 + \cdots + \eta_n\mathbf{e}_n) \\ &= (\lambda_1\xi_1\mathbf{e}_1 + \lambda_2\xi_2\mathbf{e}_2 + \cdots + \lambda_n\xi_n\mathbf{e}_n, \eta_1\mathbf{e}_1 + \eta_2\mathbf{e}_2 + \cdots + \eta_n\mathbf{e}_n) \\ &= \lambda_1\xi_1\eta_1 + \lambda_2\xi_2\eta_2 + \cdots + \lambda_n\xi_n\eta_n. \end{aligned}$$

Putting $\mathbf{y} = \mathbf{x}$ we obtain the following

THEOREM 3. *Let $A(\mathbf{x}; \mathbf{x})$ be a quadratic form on an n-dimensional Euclidean space. Then there exists an orthonormal basis relative to which the quadratic form can be represented as*

$$A(\mathbf{x}; \mathbf{x}) = \sum \lambda_i \xi_i^2.$$

Here the λ_i are the eigenvalues of the transformation A or, equivalently, the roots of the characteristic equation of the matrix $||a_{ik}||$.

For $n = 3$ the above theorem is a theorem of solid analytic geometry. Indeed, in this case the equation

$$A(\mathbf{x}; \mathbf{x}) = 1$$

is the equation of a central conic of order two. The orthonormal basis discussed in Theorem 3 defines in this case the coordinate system relative to which the surface is in canonical form. The basis vectors $\mathbf{e}_1$, $\mathbf{e}_2$, $\mathbf{e}_3$, are directed along the principal axes of the surface.

4. *Simultaneous reduction of a pair of quadratic forms to a sum of squares*

THEOREM 4. *Let $A(\mathbf{x}; \mathbf{x})$ and $B(\mathbf{x}; \mathbf{x})$ be two quadratic forms on an n-dimensional space* $\mathbf{R}$, *and let $B(\mathbf{x}; \mathbf{x})$ be positive definite. Then there exists a basis in* $\mathbf{R}$ *relative to which each form is expressed as a sum of squares.*

Proof: Let $B(\mathbf{x}; \mathbf{y})$ be the bilinear form corresponding to the quadratic form $B(\mathbf{x}; \mathbf{x})$. We define in $\mathbf{R}$ an inner product by means of the formula

$$(\mathbf{x}, \mathbf{y}) = B(\mathbf{x}; \mathbf{y}).$$

By Theorem 3 of this section there exists an orthonormal basis $\mathbf{e}_1, \mathbf{e}_2, \cdots, \mathbf{e}_n$ relative to which the form $A(\mathbf{x}; \mathbf{x})$ is expressed as a sum of squares, i.e.,

$$A(\mathbf{x}; \mathbf{x}) = \sum_{i=1}^{n} \lambda_i \xi_i^2. \tag{7}$$

Relative to an orthonormal basis an inner product takes the form

$$(\mathbf{x}, \mathbf{x}) = B(\mathbf{x}; \mathbf{x}) = \sum_{i=1}^{n} \xi_i^2. \tag{8}$$

Thus, relative to the basis $\mathbf{e}_1, \mathbf{e}_2, \cdots, \mathbf{e}_n$ each quadratic form can be expressed as a sum of squares.

5. *Orthogonal transformations*

DEFINITION. *A linear transformation* A *defined on a real n-dimensional Euclidean space is said to be orthogonal if it preserves inner products, i.e., if*

$$(\mathrm{A}\mathbf{x}, \mathrm{A}\mathbf{y}) = (\mathbf{x}, \mathbf{y}) \tag{9}$$

for all $\mathbf{x}, \mathbf{y} \in \mathbf{R}$.

Putting $\mathbf{x} = \mathbf{y}$ in (9) we get

$$|\mathrm{A}\mathbf{x}|^2 = |\mathbf{x}|^2, \tag{10}$$

that is, *an orthogonal transformation is length preserving.*

EXERCISE. Prove that condition (10) is sufficient for a transformation to be orthogonal.

Since

$$\cos \varphi = \frac{(\mathbf{x}, \mathbf{y})}{|\mathbf{x}|\,|\mathbf{y}|}$$

and since neither the numerator nor the denominator in the expression above is changed under an orthogonal transformation, it follows that an orthogonal transformation preserves the angle between two vectors.

Let $\mathbf{e}_1, \mathbf{e}_2, \cdots, \mathbf{e}_n$ be an orthonormal basis. Since an orthogonal transformation A preserves the angles between vectors and the length of vectors, it follows that the vectors $A\mathbf{e}_1, A\mathbf{e}_2, \cdots, A\mathbf{e}_n$ likewise form an orthonormal basis, i.e.,

$$(A\mathbf{e}_i, A\mathbf{e}_k) = \begin{cases} 1 & \text{for } i = k \\ 0 & \text{for } i \neq k. \end{cases} \tag{11}$$

Now let $||a_{ik}||$ be the matrix of A relative to the basis $\mathbf{e}_1, \mathbf{e}_2, \cdots, \mathbf{e}_n$. Since the columns of this matrix are the coordinates of the vectors $A\mathbf{e}_i$, conditions (11) can be rewritten as follows:

$$\sum_{\alpha=1}^{n} a_{\alpha i} a_{\alpha k} = \begin{cases} 1 & \text{for } i = k \\ 0 & \text{for } i \neq k. \end{cases} \tag{12}$$

EXERCISE. Show that conditions (11) and, consequently, conditions (12) are sufficient for a transformation to be orthogonal.

Conditions (12) can be written in matrix form. Indeed, $\sum_{\alpha=1}^{n} a_{\alpha i} a_{\alpha k}$ are the elements of the product of the transpose of the matrix of A by the matrix of A. Conditions (12) imply that *this product is the unit matrix.* Since the determinant of the product of two matrices is equal to the product of the determinants, it follows that the square of the determinant of a matrix of an orthogonal transformation is equal to one, i.e., *the determinant of a matrix of an orthogonal transformation is equal to* ± 1.

An orthogonal transformation whose determinant is equal to $+1$ is called *a proper orthogonal transformation,* whereas an orthogonal transformation whose determinant is equal to -1 is called *improper.*

EXERCISE. Show that the product of two proper or two improper orthogonal transformations is a proper orthogonal transformation and the product of a proper by an improper orthogonal transformation is an improper orthogonal transformation.

NOTE: What motivates the division of orthogonal transformations into proper and improper transformations is the fact that any orthogonal transformation which can be obtained by continuous deformation from the identity transformation is necessarily proper. Indeed, let A_t be an orthogonal transformation which depends continuously on the parameter t (this means that the elements of the matrix of the transformation relative to some basis are continuous functions of t) and let $A_0 = E$. Then the determinant of this transformation is also a continuous function of t. Since a continuous function which assumes the values ± 1 only is a constant and since for $t = 0$ the determinant of A_0 is equal to 1, it follows that for $t \neq 0$ the determinant of the transformation is equal to 1. Making use of Theorem 5 of this section one can also prove the converse, namely, that every proper orthogonal transformation can be obtained by continuous deformation of the identity transformation.

We now turn to a discussion of orthogonal transformations in one-dimensional and two-dimensional vector spaces. In the sequel we shall show that the study of orthogonal transformations in a space of arbitrary dimension can be reduced to the study of these two simpler cases.

Let $\mathbf{e}$ be a vector generating a one-dimensional space and A an orthogonal transformation defined on that space. Then $A\mathbf{e} = \lambda\mathbf{e}$ and since $(A\mathbf{e}, A\mathbf{e}) = (\mathbf{e}, \mathbf{e})$, we have $\lambda^2(\mathbf{e}, \mathbf{e}) = (\mathbf{e}, \mathbf{e})$, i.e., $\lambda = \pm 1$.

Thus we see that in a one-dimensional vector space there exist two orthogonal transformations only: the transformation $A\mathbf{x} \equiv \mathbf{x}$ and the transformation $A\mathbf{x} \equiv -\mathbf{x}$. The first is a proper and the second an improper transformation.

Now, consider an orthogonal transformation A on a two-dimensional vector space $\mathbf{R}$. Let $\mathbf{e}_1, \mathbf{e}_2$ be an orthonormal basis in $\mathbf{R}$ and let

$$\begin{bmatrix} \alpha & \beta \\ \gamma & \delta \end{bmatrix} \tag{13}$$

be the matrix of A relative to that basis.

We first study the case when A is a proper orthogonal transformation, i.e., we assume that $\alpha\delta - \beta\gamma = 1$.

The orthogonality condition implies that the product of the matrix (13) by its transpose is equal to the unit matrix, i.e., that

$$\begin{bmatrix} \alpha & \beta \\ \gamma & \delta \end{bmatrix}^{-1} = \begin{bmatrix} \alpha & \gamma \\ \beta & \delta \end{bmatrix}. \tag{14}$$

Since the determinant of the matrix (13) is equal to one, we have

$$(15)\qquad \begin{bmatrix} \alpha & \beta \\ \gamma & \delta \end{bmatrix}^{-1} = \begin{bmatrix} \delta & -\beta \\ -\gamma & \alpha \end{bmatrix}.$$

It follows from (14) and (15) that in this case the matrix of the transformation is

$$\begin{bmatrix} \alpha & -\beta \\ \beta & \alpha \end{bmatrix},$$

where $\alpha^2 + \beta^2 = 1$. Putting $\alpha = \cos \varphi$, $\beta = \sin \varphi$ we find that *the matrix of a proper orthogonal transformation on a two dimensional space relative to an orthogonal basis is of the form*

$$\begin{bmatrix} \cos \varphi & -\sin \varphi \\ \sin \varphi & \cos \varphi \end{bmatrix}$$

(a rotation of the plane by an angle φ).

Assume now that A is an improper orthogonal transformation, that is, that $\alpha\delta - \beta\gamma = -1$. In this case the characteristic equation of the matrix (13) is $\lambda^2 - (\alpha + \delta)\lambda - 1 = 0$ and, thus, has real roots. This means that the transformation A has an eigenvector $\mathbf{e}$, $A\mathbf{e} = \lambda\mathbf{e}$. Since A is orthogonal it follows that $A\mathbf{e} = \pm\mathbf{e}$. Furthermore, an orthogonal transformation preserves the angles between vectors and their length. Therefore any vector $\mathbf{e}_1$ orthogonal to $\mathbf{e}$ is transformed by A into a vector orthogonal to $A\mathbf{e} = \pm\mathbf{e}$, i.e., $A\mathbf{e}_1 = \pm\mathbf{e}_1$. Hence the matrix of A relative to the basis $\mathbf{e}$, $\mathbf{e}_1$ has the form

$$\begin{bmatrix} \pm 1 & 0 \\ 0 & \pm 1 \end{bmatrix}.$$

Since the determinant of an improper transformation is equal to -1, the canonical form of the matrix of an improper orthogonal transformation in two-dimensional space is

$$\begin{bmatrix} +1 & 0 \\ 0 & -1 \end{bmatrix} \quad \text{or} \quad \begin{bmatrix} -1 & 0 \\ 0 & +1 \end{bmatrix}$$

(a reflection in one of the axes).

We now find the simplest form of the matrix of an orthogonal transformation defined on a space of arbitrary dimension.

THEOREM 5. *Let* A *be an orthogonal transformation defined on an n-dimensional Euclidean space* **R**. *Then there exists an orthonormal basis* $\mathbf{e}_1, \mathbf{e}_2, \cdots, \mathbf{e}_n$ *of* **R** *relative to which the matrix of the transformation is*

$$\begin{bmatrix} 1 & & & & & & & & & & & & \\ & \cdot & & & & & & & & & & & \\ & & \cdot & & & & & & & & & & \\ & & & 1 & & & & & & & & & \\ & & & & -1 & & & & & & & & \\ & & & & & \cdot & & & & & & & \\ & & & & & & \cdot & & & & & & \\ & & & & & & & -1 & & & & & \\ & & & & & & & & \cos\varphi_1 & -\sin\varphi_1 & & & \\ & & & & & & & & \sin\varphi_1 & \cos\varphi_1 & & & \\ & & & & & & & & & & \cdot & & \\ & & & & & & & & & & & \cos\varphi_k & -\sin\varphi_k \\ & & & & & & & & & & & \sin\varphi_k & \cos\varphi_k \end{bmatrix}$$

where the unspecified entries have value zero.

Proof: According to Theorem 1 of this section **R** contains a one-or two-dimensional invariant subspace $\mathbf{R}^{(1)}$. If there exists a one-dimensional invariant subspace $\mathbf{R}^{(1)}$ we denote by $\mathbf{e}_1$ a vector of length one in that space. Otherwise $\mathbf{R}^{(1)}$ is two dimensional and we choose in it an orthonormal basis $\mathbf{e}_1, \mathbf{e}_2$. Consider A on $\mathbf{R}^{(1)}$. In the case when $\mathbf{R}^{(1)}$ is one-dimensional, A takes the form $A\mathbf{x} = \pm\mathbf{x}$. If $\mathbf{R}^{(1)}$ is two dimensional A is a proper orthogonal transformation (otherwise $\mathbf{R}^{(1)}$ would contain a one-dimensional invariant subspace) and the matrix of A in $\mathbf{R}^{(1)}$ is of the form

$$\begin{bmatrix} \cos\varphi & -\sin\varphi \\ \sin\varphi & \cos\varphi \end{bmatrix}.$$

The totality $\tilde{\mathbf{R}}$ of vectors orthogonal to all the vectors of $\mathbf{R}^{(1)}$ forms an invariant subspace.

Indeed, consider the case when $\mathbf{R}^{(1)}$ is a two-dimensional space, say. Let $\mathbf{x} \in \tilde{\mathbf{R}}$, i.e.,

$$(\mathbf{x}, \mathbf{y}) = 0 \quad \text{for} \quad \text{all } \mathbf{y} \in \mathbf{R}^{(1)}.$$

Since $(A\mathbf{x}, A\mathbf{y}) = (\mathbf{x}, \mathbf{y})$, it follows that $(A\mathbf{x}, A\mathbf{y}) = 0$. As $\mathbf{y}$ varies over all of $\mathbf{R}^{(1)}$, $\mathbf{z} = A\mathbf{y}$ likewise varies over all of $\mathbf{R}^{(1)}$. Hence $(A\mathbf{x}, \mathbf{z}) = 0$ for all $\mathbf{z} \in \mathbf{R}^{(1)}$, i.e., $A\mathbf{x} \in \tilde{\mathbf{R}}$. We reason analogously if $\mathbf{R}^{(1)}$ is one-dimensional. If $\mathbf{R}^{(1)}$ is of dimension one, $\tilde{\mathbf{R}}$ is of dimension $n - 1$. Again, if $\mathbf{R}^{(1)}$ is of dimension two, $\tilde{\mathbf{R}}$ is of dimension $n - 2$. Indeed, in the former case, $\tilde{\mathbf{R}}$ is the totality of vectors orthogonal to the vector $\mathbf{e}_1$, and in the latter case, $\tilde{\mathbf{R}}$ is the totality of vectors orthogonal to the vectors $\mathbf{e}_1$ and $\mathbf{e}_2$.

We now find a one-dimensional or two-dimensional invariant subspace of $\tilde{\mathbf{R}}$, select a basis in it, etc.

In this manner we obtain n pairwise orthogonal vectors of length one which form a basis of $\mathbf{R}$. Relative to this basis the matrix of the transformation is of the form

$$\begin{bmatrix} 1 & & & & & & & & & & & & \\ & \cdot & & & & & & & & & & & \\ & & \cdot & & & & & & & & & & \\ & & & \cdot & & & & & & & & & \\ & & & & 1 & & & & & & & & \\ & & & & & -1 & & & & & & & \\ & & & & & & \cdot & & & & & & \\ & & & & & & & \cdot & & & & & \\ & & & & & & & & \cdot & & & & \\ & & & & & & & & & -1 & & & \\ & & & & & & & & & & \cos\varphi_1 & -\sin\varphi_1 & & & \\ & & & & & & & & & & \sin\varphi_1 & \cos\varphi_1 & & & \\ & & & & & & & & & & & & \cdot & & \\ & & & & & & & & & & & & & \cdot & \\ & & & & & & & & & & & & & & \cdot \\ & & & & & & & & & & & & & & & \cos\varphi_k & -\sin\varphi_k \\ & & & & & & & & & & & & & & & \sin\varphi_k & \cos\varphi_k \end{bmatrix},$$

where the ± 1 on the principal diagonal correspond to one-dimensional invariant subspaces and the "boxes"

$$\begin{bmatrix} \cos\varphi_i & -\sin\varphi_i \\ \sin\varphi_i & \cos\varphi_i \end{bmatrix}$$

correspond to two-dimensional invariant subspaces. This completes the proof of the theorem.

NOTE: A proper orthogonal transformation which represents a rotation of a two-dimensional plane and which leaves the $(n-2)$-dimensional subspace orthogonal to that plane fixed is called *a simple rotation*. Relative to a suitable basis its matrix is of the form

$$\begin{bmatrix} 1 & & & & & & & & \\ & \cdot & & & & & & & \\ & & \cdot & & & & & & \\ & & & 1 & & & & & \\ & & & & \cos\varphi & -\sin\varphi & & & \\ & & & & \sin\varphi & \cos\varphi & & & \\ & & & & & & 1 & & \\ & & & & & & & \cdot & \\ & & & & & & & & 1 \end{bmatrix}.$$

An improper orthogonal transformation which reverses all vectors of some one-dimensional subspace and leaves all the vectors of the $(n-1)$-dimensional complement fixed is called *a simple reflection*. Relative to a suitable basis its matrix takes the form

$$\begin{bmatrix} 1 & & & & & & \\ & \cdot & & & & & \\ & & 1 & & & & \\ & & & -1 & & & \\ & & & & 1 & & \\ & & & & & \cdot & \\ & & & & & & 1 \end{bmatrix}.$$

Making use of Theorem 5 one can easily show that every orthogonal transformation can be written as the product of a number of simple rotations and simple reflections. The proof is left to the reader.

§ 17. Extremal properties of eigenvalues

In this section we show that the eigenvalues of a self-adjoint linear transformation defined on an n-dimensional Euclidean space can be obtained by considering a certain minimum problem connected with the corresponding quadratic form $(\mathbf{Ax}, \mathbf{x})$. This approach will, in particular, permit us to prove the existence of eigenvalues and eigenvectors without making use of the theorem on the existence of a root of an nth order equation. The extremal properties are also useful in computing eigenvalues. We shall first consider the case of a real space and then extend our results to the case of a complex space.

We first prove the following lemma:

LEMMA 1. *Let* B *be a self-adjoint linear transformation on a real space such that the quadratic form* $(\mathrm{B}\mathbf{x}, \mathbf{x})$ *is non-negative, i.e., such that*

$$(\mathrm{B}\mathbf{x}, \mathbf{x}) \geqq 0 \qquad \textit{for all } \mathbf{x}.$$

If for some vector $\mathbf{x} = \mathbf{e}$

$$(\mathrm{B}\mathbf{e}, \mathbf{e}) = 0,$$

then $\mathrm{B}\mathbf{e} = 0$.

Proof: Let $\mathbf{x} = \mathbf{e} + t\mathbf{h}$, where t is an arbitrary number and $\mathbf{h}$ a vector. We have

$$(\mathrm{B}(\mathbf{e} + t\mathbf{h}), \mathbf{e} + t\mathbf{h}) = (\mathrm{B}\mathbf{e}, \mathbf{e}) + t(\mathrm{B}\mathbf{e}, \mathbf{h}) + t(\mathrm{B}\mathbf{h}, \mathbf{e}) + t^2(\mathrm{B}\mathbf{h}, \mathbf{h}) \geqq 0.$$

Since $(\mathrm{B}\mathbf{h}, \mathbf{e}) = (\mathbf{h}, \mathrm{B}\mathbf{e}) = (\mathrm{B}\mathbf{e}, \mathbf{h})$ and $(\mathrm{B}\mathbf{e}, \mathbf{e}) = 0$, then $2t(\mathrm{B}\mathbf{e}, \mathbf{h}) + t^2(\mathrm{B}\mathbf{h}, \mathbf{h}) \geqq 0$ for all t. But this means that $(\mathrm{B}\mathbf{e}, \mathbf{h}) = 0$.

Indeed, the function $at + bt^2$ with $a \neq 0$ changes sign at $t = 0$. However, in our case the expression

$$2t(\mathrm{B}\mathbf{e}, \mathbf{h}) + t^2(\mathrm{B}\mathbf{h}, \mathbf{h})$$

is non-negative for all t. It follows that

$$(\mathrm{B}\mathbf{e}, \mathbf{h}) = 0.$$

Since $\mathbf{h}$ was arbitrary, $\mathrm{B}\mathbf{e} = 0$. This proves the lemma.

Let A be a self-adjoint linear transformation on an n-dimensional real Euclidean space. We shall consider the quadratic form $(\mathrm{A}\mathbf{x}, \mathbf{x})$ which corresponds to A on the *unit sphere*, i.e., on the set of vectors $\mathbf{x}$ such that

$$(\mathbf{x}, \mathbf{x}) = 1.$$

THEOREM 1. *Let* A *be a self-adjoint linear transformation. Then the quadratic form* $(\mathrm{A}\mathbf{x}, \mathbf{x})$ *corresponding to* A *assumes its minimum* λ_1 *on the unit sphere. The vector* $\mathbf{e}_1$ *at which the minimum is assumed is an eigenvector of* A *and* λ_1 *is the corresponding eigenvalue.*

Proof: The unit sphere is a closed and bounded set in n-dimensional space. Since $(A\mathbf{x}, \mathbf{x})$ is continuous on that set it must assume its minimum λ_1 at some point $\mathbf{e}_1$. We have

$$(1) \qquad (A\mathbf{x}, \mathbf{x}) \geqq \lambda_1 \qquad \text{for } (\mathbf{x}, \mathbf{x}) = 1,$$

and

$$(A\mathbf{e}_1, \mathbf{e}_1) = \lambda_1, \qquad \text{where } (\mathbf{e}_1, \mathbf{e}_1) = 1.$$

Inequality (1) can be rewritten as follows

$$(2) \qquad (A\mathbf{x}, \mathbf{x}) \geqq \lambda_1(\mathbf{x}, \mathbf{x}), \qquad \text{where } (\mathbf{x}, \mathbf{x}) = 1.$$

This inequality holds for vectors of unit length. Note that if we multiply $\mathbf{x}$ by some number α, then both sides of the inequality become multiplied by α^2. Since any vector can be obtained from a vector of unit length by multiplying it by some number α, it follows that inequality (2) holds for vectors of arbitrary length.

We now rewrite (2) in the form

$$(A\mathbf{x} - \lambda_1\mathbf{x}, \mathbf{x}) \geqq 0 \qquad \text{for all } \mathbf{x}.$$

In particular, for $\mathbf{x} = \mathbf{e}_1$, we have

$$(A\mathbf{e}_1 - \lambda_1\mathbf{e}_1, \mathbf{e}) = 0.$$

This means that the transformation $B = A - \lambda_1 E$ satisfies the conditions of Lemma 1. Hence

$$(A - \lambda_1 E)\mathbf{e}_1 = \mathbf{0}, \text{ i.e., } A\mathbf{e}_1 = \lambda_1\mathbf{e}_1.$$

We have shown that $\mathbf{e}_1$ is an eigenvector of the transformation A corresponding to the eigenvalue λ_1. This proves the theorem.

To find the next eigenvalue of A we consider all vectors of $\mathbf{R}$ orthogonal to the eigenvector $\mathbf{e}_1$. As was shown in para. 2, § 16 (Lemma 2), these vectors form an $(n - 1)$-dimensional subspace $\mathbf{R}_1$ invariant under A. The required second eigenvalue λ_2 of A is the minimum of $(A\mathbf{x}, \mathbf{x})$ on the unit sphere in $\mathbf{R}_1$. The corresponding eigenvector $\mathbf{e}_2$ is the point in $\mathbf{R}_1$ at which the minimum is assumed.

Obviously, $\lambda_2 \geqq \lambda_1$ since the minimum of a function considered on the whole space cannot exceed the minimum of the function in a subspace.

We obtain the next eigenvector by solving the same problem in the $(n-2)$-dimensional subspace consisting of vectors orthogonal to both $\mathbf{e}_1$ and $\mathbf{e}_2$. The third eigenvalue of A is equal to the minimum of $(A\mathbf{x}, \mathbf{x})$ in that subspace.

Continuing in this manner we find all the n eigenvalues and the corresponding eigenvectors of A.

It is sometimes convenient to determine the second, third, etc., eigenvector of a transformation from the extremum problem without reference to the preceding eigenvectors.

Let A be a self-adjoint transformation. Denote by

$$\lambda_1 \leqq \lambda_2 \leqq \cdots \leqq \lambda_n$$

its eigenvalues and by $\mathbf{e}_1, \mathbf{e}_2, \cdots, \mathbf{e}_n$ the corresponding orthonormal eigenvectors.

We shall show that *if* $\mathbf{S}$ *is the subspace spanned by the first* k *eigenvectors*

$$\mathbf{e}_1, \mathbf{e}_2, \cdots, \mathbf{e}_k$$

then for each $\mathbf{x} \in \mathbf{S}$ *the following inequality holds*:

$$\lambda_1(\mathbf{x}, \mathbf{x}) \leqq (A\mathbf{x}, \mathbf{x}) \leqq \lambda_k(\mathbf{x}, \mathbf{x}).$$

Indeed, let

$$\mathbf{x} = \xi_1\mathbf{e}_1 + \xi_2\mathbf{e}_2 + \cdots + \xi_k\mathbf{e}_k.$$

Since $A\mathbf{e}_k = \lambda_k\mathbf{e}_k$, $(\mathbf{e}_k, \mathbf{e}_k) = 1$ and $(\mathbf{e}_k, \mathbf{e}_i) = 0$ for $i \neq k$, it follows that

$$\begin{aligned}(A\mathbf{x}, \mathbf{x}) &= (A(\xi_1\mathbf{e}_1 + \xi_2\mathbf{e}_2 + \cdots + \xi_k\mathbf{e}_k), \xi_1\mathbf{e}_1 + \xi_2\mathbf{e}_2 + \cdots + \xi_k\mathbf{e}_k) \\ &= (\lambda_1\xi_1\mathbf{e}_1 + \lambda_2\xi_2\mathbf{e}_2 + \cdots + \lambda_k\xi_k\mathbf{e}_k, \xi_1\mathbf{e}_1 + \xi_2\mathbf{e}_2 + \cdots + \xi_k\mathbf{e}_k) \\ &= \lambda_1\xi_1^2 + \lambda_2\xi_2^2 + \cdots + \lambda_k\xi_k^2.\end{aligned}$$

Furthermore, since $\mathbf{e}_1, \mathbf{e}_2, \cdots, \mathbf{e}_k$ are orthonormal,

$$(\mathbf{x}, \mathbf{x}) = \xi_1^2 + \xi_2^2 + \cdots + \xi_k^2$$

and therefore

$$(A\mathbf{x}, \mathbf{x}) = \lambda_1\xi_1^2 + \lambda_2\xi_2^2 + \cdots + \lambda_k\xi_k^2 \geqq \lambda_1(\xi_1^2 + \xi_2^2 + \cdots + \xi_k^2) = \\ = \lambda_1(\mathbf{x}, \mathbf{x}).$$

Similarly,

$$(A\mathbf{x}, \mathbf{x}) \leqq \lambda_k(\mathbf{x}, \mathbf{x}).$$

It follows that

$$\lambda_1(\mathbf{x}, \mathbf{x}) \leqq (A\mathbf{x}, \mathbf{x}) \leqq \lambda_k(\mathbf{x}, \mathbf{x}).$$

Now let $\mathbf{R}_k$ be a subspace of dimension $n - k + 1$. In § 7 (Lemma of para. 1) we showed that if the sum of the dimensions of two subspaces of an n-dimensional space is greater than n, then there exists a vector different from zero belonging to both subspaces. Since the sum of the dimensions of $\mathbf{R}_k$ and $\mathbf{S}$ is $(n - k + 1) + k$ it follows that there exists a vector $\mathbf{x}_0$ common to both $\mathbf{R}_k$ and $\mathbf{S}$. We can assume that $\mathbf{x}_0$ has unit length, that is,

$(\mathbf{x}_0, \mathbf{x}_0) = 1$. Since $(A\mathbf{x}, \mathbf{x}) \leqq \lambda_k (\mathbf{x}, \mathbf{x})$ for $\mathbf{x} \in \mathbf{S}$, it follows that $(A\mathbf{x}_0, \mathbf{x}_0) \leqq \lambda_k$.

We have thus shown that there exists a vector $\mathbf{x}_0 \in \mathbf{R}_k$ of unit length such that

$$(A\mathbf{x}_0, \mathbf{x}_0) \leqq \lambda_k.$$

But then the minimum of $(A\mathbf{x}, \mathbf{x})$ for $\mathbf{x}$ on the unit sphere in $\mathbf{R}_k$ must be equal to or less than λ_k.

To sum up: *If $\mathbf{R}_k$ is an $(n - k + 1)$-dimensional subspace and $\mathbf{x}$ varies over all vectors in $\mathbf{R}_k$ for which $(\mathbf{x}, \mathbf{x}) = 1$, then*

$$\min (A\mathbf{x}, \mathbf{x}) \leqq \lambda_k.$$

Note that among all the subspaces of dimension $n - k + 1$ there exists one for which $\min (A\mathbf{x}, \mathbf{x})$, $(\mathbf{x}, \mathbf{x}) = 1$, $\mathbf{x} \in \mathbf{R}_k$, is actually equal to λ_k. This is the subspace consisting of all vectors orthogonal to the first k eigenvectors $\mathbf{e}_1, \mathbf{e}_2, \cdots, \mathbf{e}_k$. Indeed, we showed in this section that min $(A\mathbf{x}, \mathbf{x})$, $(\mathbf{x}, \mathbf{x}) = 1$, taken over all vectors orthogonal to $\mathbf{e}_1, \mathbf{e}_2, \cdots, \mathbf{e}_k$ is equal to λ_k.

We have thus proved the following theorem:

THEOREM. *Let $\mathbf{R}_k$ be a $(n - k + 1)$-dimensional subspace of the space $\mathbf{R}$. Then* $\min (A\mathbf{x}, \mathbf{x})$ *for all $\mathbf{x} \in \mathbf{R}_k$, $(\mathbf{x}, \mathbf{x}) = 1$, is less than or equal to λ_k. The subspace $\mathbf{R}_k$ can be chosen so that* $\min(A\mathbf{x}, \mathbf{x})$ *is equal to λ_k.*

Our theorem can be expressed by the formula

$$\max_{\mathbf{R}_k} \min_{\substack{(\mathbf{x}, \mathbf{x})=1 \\ \mathbf{x} \in \mathbf{R}_k}} (A\mathbf{x}, \mathbf{x}) = \lambda_k. \tag{3}$$

In this formula the minimum is taken over all $\mathbf{x} \in \mathbf{R}_k$, $(\mathbf{x}, \mathbf{x}) = 1$, and the maximum over all subspaces $\mathbf{R}_k$ of dimension $n - k + 1$.

As a consequence of our theorem we have:

Let A be a self-adjoint linear transformation and B a postive definite linear transformation. Let $\lambda_1 \leqq \lambda_2 \leqq \cdots \leqq \lambda_n$ be the eigenvalues of A and let $\mu_1 \leqq \mu_2 \leqq \cdots \leqq \mu_n$ be the eigenvalues of A + B. Then $\lambda_k \leqq \mu_k$.

Indeed

$$(A\mathbf{x}, \mathbf{x}) \leqq ((A + B)\mathbf{x}, \mathbf{x}),$$

for all $\mathbf{x}$. Hence for any $(n - k + 1)$-dimensional subspace $\mathbf{R}_k$ we have

$$\min_{\substack{(\mathbf{x}, \mathbf{x})=1 \\ \mathbf{x} \in \mathbf{R}_k}} (A\mathbf{x}, \mathbf{x}) \leqq \min_{\substack{(\mathbf{x}, \mathbf{x})=1 \\ \mathbf{x} \in \mathbf{R}_k}} ((A + B)\mathbf{x}, \mathbf{x}).$$

It follows that the maximum of the expression on the left side taken over all subspaces $\mathbf{R}_k$ does not exceed the maximum of the right side. Since, by formula (3), the maximum of the left side is equal to λ_k and the maximum of the right side is equal to μ_k, we have $\lambda_k \leqq \mu_k$.

We now extend our results to the case of a complex space.

To this end we need only substitute for Lemma 1 the following lemma.

LEMMA 2. *Let* B *be a self-adjoint transformation on a complex space and let the Hermitian form* $(\mathrm{B}\mathbf{x}, \mathbf{x})$ *corresponding to* B *be non-negative, i.e., let*

$$(\mathrm{B}\mathbf{x}, \mathbf{x}) \geqq 0 \qquad \textit{for all } \mathbf{x}.$$

If for some vector $\mathbf{e}$, $(\mathrm{B}\mathbf{e}, \mathbf{e}) = 0$, *then* $\mathrm{B}\mathbf{e} = 0$.

Proof: Let t be an arbitrary real number and $\mathbf{h}$ a vector. Then

$$(\mathrm{B}(\mathbf{e} + t\mathbf{h}),\ \mathbf{e} + t\mathbf{h}) \geqq 0,$$

or, since $(\mathrm{B}\mathbf{e}, \mathbf{e}) = 0$,

$$t[(\mathrm{B}\mathbf{e}, \mathbf{h}) + (\mathrm{B}\mathbf{h}, \mathbf{e})] + t^2(\mathrm{B}\mathbf{h}, \mathbf{h}) \geqq 0$$

for all t. It follows that

$$(\mathrm{B}\mathbf{e}, \mathbf{h}) + (\mathrm{B}\mathbf{h}, \mathbf{e}) = 0. \tag{4}$$

Since $\mathbf{h}$ was arbitrary, we get, by putting $i\mathbf{h}$ in place of $\mathbf{h}$,

$$-i(\mathrm{B}\mathbf{e}, \mathbf{h}) + i(\mathrm{B}\mathbf{h}, \mathbf{e}) = 0. \tag{5}$$

It follows from (4) and (5) that

$$(\mathrm{B}\mathbf{e}, \mathbf{h}) = 0,$$

and therefore $\mathrm{B}\mathbf{e} = 0$. This proves the lemma.

All the remaining results of this section as well as their proofs can be carried over to complex spaces without change.

CHAPTER III

The Canonical Form of an Arbitrary Linear Transformation

§ 18. The canonical form of a linear transformation

In chapter II we discussed various classes of linear transformations on an n-dimensional vector space which have n linearly independent eigenvectors. We found that relative to the basis consisting of the eigenvectors the matrix of such a transformation had a particularly simple form, namely, the so-called diagonal form.

However, the number of linearly independent eigenvectors of a linear transformation can be less than n. [1] (An example of such a transformation is given in the sequel; cf. also § 10, para. 1, Example 3). Clearly, such a transformation is not diagonable since, as noted above, any basis relative to which the matrix of a transformation is diagonal consists of linearly independent eigenvectors of the transformation. There arises the question of the simplest form of such a transformation.

In this chapter we shall find for an arbitrary transformation a basis relative to which the matrix of the transformation has a comparatively simple form (the so-called *Jordan canonical form*). In the case when the number of linearly independent eigenvectors of the transformation is equal to the dimension of the space the canonical form will coincide with the diagonal form. We now formulate the definitive result which we shall prove in § 19.

Let A *be an arbitrary linear transformation on a complex n-dimensional space and let* A *have k $(k \leqq n)$ linearly independent eigenvectors*

[1] We recall that if the characteristic polynomial has n distinct roots, then the transformation has n linearly independent eigenvectors. Hence for the number of linearly independent eigenvectors of a transformation to be less than n it is necessary that the characteristic polynomial have multiple roots. Thus, this case is, in a sense, exceptional.

$$\mathbf{e}_1, \mathbf{f}_1, \cdots, \mathbf{h}_1,$$

corresponding to the eigenvalues $\lambda_1, \lambda_2, \cdots, \lambda_k$. *Then there exists a basis consisting of* k *sets of vectors* [2]

$$\text{(1)} \qquad \mathbf{e}_1, \cdots, \mathbf{e}_p; \qquad \mathbf{f}_1, \cdots, \mathbf{f}_q; \qquad \cdots; \qquad \mathbf{h}_1, \cdots, \mathbf{h}_s,$$

relative to which the transformation A *has the form*:

$$\text{(2)} \qquad \begin{array}{lll} \mathrm{A}\mathbf{e}_1 = \lambda_1\mathbf{e}_1, & \mathrm{A}\mathbf{e}_2 = \mathbf{e}_1 + \lambda_1\mathbf{e}_2, \cdots, & \mathrm{A}\mathbf{e}_p = \mathbf{e}_{p-1} + \lambda_1\mathbf{e}_p; \\ \mathrm{A}\mathbf{f}_1 = \lambda_2\mathbf{f}_1, & \mathrm{A}\mathbf{f}_2 = \mathbf{f}_1 + \lambda_2\mathbf{f}_2, \cdots, & \mathrm{A}\mathbf{f}_q = \mathbf{f}_{q-1} + \lambda_2\mathbf{f}_q; \\ \cdots\cdots & \cdots\cdots & \cdots\cdots \\ \mathrm{A}\mathbf{h}_1 = \lambda_k\mathbf{h}_1, & \mathrm{A}\mathbf{h}_2 = \mathbf{h}_1 + \lambda_k\mathbf{h}_2, \cdots, & \mathrm{A}\mathbf{h}_s = \mathbf{h}_{s-1} + \lambda_k\mathbf{h}_s. \end{array}$$

We see that the linear transformation A described by (2) takes the basis vectors of each set into linear combinations of vectors in the same set. It therefore follows that each set of basis vectors generates a subspace invariant under A. We shall now investigate A more closely.

Every subspace generated by each one of the k sets of vectors contains an eigenvector. For instance, the subspace generated by the set $\mathbf{e}_1, \cdots, \mathbf{e}_p$ contains the eigenvector $\mathbf{e}_1$. We show that each subspace contains only one (to within a multiplicative constant) eigenvector. Indeed, consider the subspace generated by the vectors $\mathbf{e}_1, \mathbf{e}_2, \cdots, \mathbf{e}_p$, say. Assume that some vector of this subspace, i.e., some linear combination of the form

$$c_1\mathbf{e}_1 + c_2\mathbf{e}_2 + \cdots + c_p\mathbf{e}_p,$$

where not all the c's are equal to zero, is an eigenvector, that is,

$$\mathrm{A}(c_1\mathbf{e}_1 + c_2\mathbf{e}_2 + \cdots + c_p\mathbf{e}_p) = \lambda(c_1\mathbf{e}_1 + c_2\mathbf{e}_2 + \cdots + c_p\mathbf{e}_p).$$

Substituting the appropriate expressions of formula (2) on the left side we obtain

$$c_1\lambda_1\mathbf{e}_1 + c_2(\mathbf{e}_1 + \lambda_1\mathbf{e}_2) + \cdots + c_p(\mathbf{e}_{p-1} + \lambda_1\mathbf{e}_p) = \\ = \lambda c_1\mathbf{e}_1 + \lambda c_2\mathbf{e}_2 + \cdots + \lambda c_p\mathbf{e}_p.$$

Equating the coefficients of the basis vectors we get a system of equations for the numbers $\lambda, c_1, c_2, \cdots, c_p$:

[2] Clearly, $p + q + \cdots + s = n$. If $k = n$, then each set consists of one vector only, namely an eigenvector.

$$\begin{aligned} c_1\lambda_1 + c_2 &= \lambda c_1, \\ c_2\lambda_1 + c_3 &= \lambda c_2, \\ &\cdots\cdots\cdots \\ c_{p-1}\lambda_1 + c_p &= \lambda c_{p-1}, \\ c_p\lambda_1 \qquad &= \lambda c_p. \end{aligned}$$

We first show that $\lambda = \lambda_1$. Indeed, if $\lambda \neq \lambda_1$, then it would follow from the last equation that $c_p = 0$ and from the remaining equations that $c_{p-1} = c_{p-2} = \cdots = c_2 = c_1 = 0$. Hence $\lambda = \lambda_1$. Substituting this value for λ we get from the first equation $c_2 = 0$, from the second, $c_3 = 0, \cdots$ and from the last, $c_p = 0$. This means that the eigenvector is equal to $c_1\mathbf{e}_1$ and, therefore, coincides (to within a multiplicative constant) with the first vector of the corresponding set.

We now write down the matrix of the transformation (2). Since the vectors of each set are transformed into linear combinations of vectors of the same set, it follows that in the first p columns the row indices of possible non-zero elements are $1, 2, \cdots, p$; in the next q columns the row indices of possible non zero elements are $p + 1, p + 2, \cdots, p + q$, and so on. Thus, the matrix of the transformation relative to the basis (1) has k boxes along the main diagonal. The elements of the matrix which are outside these boxes are equal to zero.

To find out what the elements in each box are it suffices to note how A transforms the vectors of the appropriate set. We have

$$\begin{aligned} \mathrm{A}\mathbf{e}_1 &= \lambda_1\mathbf{e}_1, \\ \mathrm{A}\mathbf{e}_2 &= \mathbf{e}_1 + \lambda_1\mathbf{e}_2, \\ &\cdots\cdots\cdots \\ \mathrm{A}\mathbf{e}_{p-1} &= \qquad \mathbf{e}_{p-2} + \lambda_1\mathbf{e}_{p-1}, \\ \mathrm{A}\mathbf{e}_p &= \qquad\qquad \mathbf{e}_{p-1} + \lambda_1\mathbf{e}_p. \end{aligned}$$

Recalling how one constructs the matrix of a transformation relative to a given basis we see that the box corresponding to the set of vectors $\mathbf{e}_1, \mathbf{e}_2, \cdots, \mathbf{e}_p$ has the form

$$(3) \qquad \mathscr{A}_1 = \begin{bmatrix} \lambda_1 & 1 & 0 & \cdots & 0 & 0 \\ 0 & \lambda_1 & 1 & \cdots & 0 & 0 \\ \cdots & \cdots & \cdots & \cdots & \cdots & \cdots \\ 0 & 0 & 0 & \cdots & \lambda_1 & 1 \\ 0 & 0 & 0 & \cdots & 0 & \lambda_1 \end{bmatrix}.$$

The matrix of A consists of similar boxes of orders $p, q, \cdots, s$, that is, it has the form

$$
(4) \qquad \begin{bmatrix}
\lambda_1 & 1 & 0 & \cdots & 0 & & & & & & & & & & \\
0 & \lambda_1 & 1 & \cdots & 0 & & & & & & & & & & \\
\cdot & \cdot & \cdot & \cdot & \cdot & & & & & & & & & & \\
0 & 0 & 0 & \cdots & \lambda_1 & & & & & & & & & & \\
& & & & & \lambda_2 & 1 & 0 & \cdots & 0 & & & & & \\
& & & & & 0 & \lambda_2 & 1 & \cdots & 0 & & & & & \\
& & & & & \cdot & \cdot & \cdot & \cdot & \cdot & & & & & \\
& & & & & 0 & 0 & 0 & \cdots & \lambda_2 & & & & & \\
& & & & & & & & & & \ddots & & & & \\
& & & & & & & & & & & \lambda_k & 1 & 0 & \cdots & 0 \\
& & & & & & & & & & & 0 & \lambda_k & 1 & \cdots & 0 \\
& & & & & & & & & & & \cdot & \cdot & \cdot & \cdot & \cdot \\
& & & & & & & & & & & 0 & 0 & 0 & \cdots & \lambda_k
\end{bmatrix}.
$$

Here all the elements outside of the boxes are zero.

Although a matrix in the canonical form described above seems more complicated than a diagonal matrix, say, one can nevertheless perform algebraic operations on it with relative ease. We show, for instance, how to compute a polynomial in the matrix (4). The matrix (4) has the form

$$
\mathscr{A} = \begin{bmatrix} \mathscr{A}_1 & & & & \\ & \mathscr{A}_2 & & & \\ & & \cdot & & \\ & & & \cdot & \\ & & & & \cdot & \\ & & & & & \mathscr{A}_k \end{bmatrix},
$$

where the $\mathscr{A}_1$ are square boxes and all other elements are zero. Then

$$
\mathscr{A}^2 = \begin{bmatrix} \mathscr{A}_1^2 & & & & \\ & \mathscr{A}_2^2 & & & \\ & & \cdot & & \\ & & & \cdot & \\ & & & & \cdot & \\ & & & & & \mathscr{A}_k^2 \end{bmatrix}, \quad \cdots, \quad \mathscr{A}^m = \begin{bmatrix} \mathscr{A}_1^m & & & & \\ & \mathscr{A}_2^m & & & \\ & & \cdot & & \\ & & & \cdot & \\ & & & & \cdot & \\ & & & & & \mathscr{A}_k^m \end{bmatrix},
$$

that is, in order to raise the matrix $\mathscr{A}$ to some power all one has to do is raise each one of the boxes to that power. Now let $P(t) = a_0 + a_1 t + \cdots + a_m t^m$ be any polynomial. It is easy to see that

$$P(\mathcal{A}) = \begin{bmatrix} P(\mathcal{A}_1) & & & & \\ & P(\mathcal{A}_2) & & & \\ & & \cdot & & \\ & & & \cdot & \\ & & & & \cdot \\ & & & & & P(\mathcal{A}_k) \end{bmatrix}.$$

We now show how to compute $P(\mathcal{A}_1)$, say. First we write the matrix $\mathcal{A}_1$ in the form

$$\mathcal{A}_1 = \lambda_1 \mathcal{E} + \mathcal{I},$$

where $\mathcal{E}$ is the unit matrix of order p and where the matrix $\mathcal{I}$ has the form

$$\mathcal{I} = \begin{bmatrix} 0 & 1 & 0 & \cdots & 0 & 0 \\ 0 & 0 & 1 & \cdots & 0 & 0 \\ 0 & 0 & 0 & \cdots & 0 & 1 \\ 0 & 0 & 0 & \cdots & 0 & 0 \end{bmatrix}.$$

We note that the matrices $\mathcal{I}^2, \mathcal{I}^3, \cdots, \mathcal{I}^{p-1}$ are of the form [2]

$$\mathcal{I}^2 = \begin{bmatrix} 0 & 0 & 1 & 0 & \cdots & 0 \\ 0 & 0 & 0 & 1 & \cdots & 0 \\ \cdots & \cdots & \cdots & \cdots & \cdots & \cdots \\ 0 & 0 & 0 & 0 & \cdots & 0 \\ 0 & 0 & 0 & 0 & \cdots & 0 \end{bmatrix}, \qquad \cdots, \qquad \mathcal{I}^{p-1} = \begin{bmatrix} 0 & 0 & 0 & \cdots & 0 & 1 \\ 0 & 0 & 0 & \cdots & 0 & 0 \\ \cdots & \cdots & \cdots & \cdots & \cdots & \cdots \\ 0 & 0 & 0 & \cdots & 0 & 0 \\ 0 & 0 & 0 & \cdots & 0 & 0 \end{bmatrix}$$

and

$$\mathcal{I}^p = \mathcal{I}^{p+1} = \cdots = 0.$$

It is now easy to compute $P(\mathcal{A}_1)$. In view of Taylor's formula a polynomial $P(t)$ can be written as

$$P(t) = P(\lambda_1) + (t - \lambda_1)P'(\lambda_1) + \frac{(t-\lambda_1)^2}{2!} P''(\lambda_1) + \cdots + \frac{(t-\lambda_1)^n}{n!} P^{(n)}(\lambda_1),$$

where n is the degree of $P(t)$. Substituting for t the matrix $\mathcal{A}_1$ we get

$$P(\mathcal{A}_1) = P(\lambda_1)\mathcal{E} + (\mathcal{A}_1 - \lambda_1 \mathcal{E})P'(\lambda_1) + \frac{(\mathcal{A}_1 - \lambda_1 \mathcal{E})^2}{2!} P''(\lambda_1) + \cdots + \frac{(\mathcal{A}_1 - \lambda_1 \mathcal{E})^n}{n!} P^{(n)}(\lambda_1).$$

But $\mathcal{A}_1 - \lambda_1 \mathcal{E} = \mathcal{I}$. Hence

$$P(\mathcal{A}_1) = P(\lambda_1)\mathcal{E} + P'(\lambda_1)\mathcal{I} + \frac{P''(\lambda_1)}{2!}\mathcal{I}^2 + \cdots + \frac{P^{(n)}(\lambda_1)}{n!}\mathcal{I}^n.$$

[2] The powers of the matrix $\mathcal{I}$ are most easily computed by observing that $\mathcal{I}\mathbf{e}_1 = 0$, $\mathcal{I}\mathbf{e}_2 = \mathbf{e}_1, \cdots, \mathcal{I}\mathbf{e}_p = \mathbf{e}_{p-1}$. Hence $\mathcal{I}^2\mathbf{e}_1 = 0$, $\mathcal{I}^2\mathbf{e}_2 = 0$, $\mathcal{I}^2\mathbf{e}_3 = \mathbf{e}_1, \cdots, \mathcal{I}^2\mathbf{e}_p = \mathbf{e}_{p-2}$. Similarly, $\mathcal{I}^3\mathbf{e}_1 = \mathcal{I}^3\mathbf{e}_2 = \mathcal{I}^3\mathbf{e}_3 = 0$, $\mathcal{I}^3\mathbf{e}_4 = \mathbf{e}_1, \cdots, \mathcal{I}^3\mathbf{e}_p = \mathbf{e}_{p-3}$.

Recalling that $\mathscr{I}^p = \mathscr{I}^{p-1} = \cdots = 0$, we get

$$P(\mathscr{A}_1) = \begin{bmatrix} P(\lambda_1) & \dfrac{P'(\lambda_1)}{1!} & \dfrac{P''(\lambda_1)}{2!} & \cdots & \dfrac{P^{(p-1)}(\lambda_1)}{(p-1)!} \\ 0 & P(\lambda_1) & \dfrac{P'(\lambda_1)}{1!} & \cdots & \dfrac{P^{(p-2)}(\lambda_1)}{(p-2)!} \\ \cdots & \cdots & \cdots & \cdots & \cdots \\ 0 & 0 & 0 & \cdots & P(\lambda_1) \end{bmatrix}.$$

Thus in order to compute $P(\mathscr{A}_1)$ where $\mathscr{A}_1$ has order p it suffices to know the value of $P(t)$ and its first $p - 1$ derivatives at the point λ_1, where λ_1 is the eigenvalue of $\mathscr{A}_1$. It follows that if the matrix has canonical form (4) with boxes of order $p, q, \cdots, s$, then to compute $P(\mathscr{A})$ one has to know the value of $P(t)$ at the points $t = \lambda_1, \lambda_2, \cdots, \lambda_k$ as well as the values of the first $p - 1$ derivatives at λ_1, the first $q - 1$ derivatives at $\lambda_2, \cdots$, and the first $s - 1$ derivatives at λ_k.

§ 19. *Reduction to canonical form*

In this section we prove the following theorem[3]:

THEOREM 1. *Let* A *be a linear transformation on a complex n-dimensional space. Then there exists a basis relative to which the matrix of the linear transformation has canonical form. In other words, there exists a basis relative to which* A *has the form* (2) (§ 18).

We prove the theorem by induction, i.e., we assume that the required basis exists in a space of dimension n and show that such a basis exists in a space of dimension $n + 1$. We need the following lemma:

LEMMA. *Every linear transformation* A *on an n-dimensional complex space* **R** *has at least one* $(n - 1)$*-dimensional invariant subspace* **R**$'$.

Proof: Consider the adjoint A* of A. Let **e** be an eigenvector of A*,

$$\mathrm{A}^*\mathbf{e} = \lambda\mathbf{e}.$$

We claim that the $(n - 1)$-dimensional subspace **R**$'$ consisting of

[3] The main idea for the proof of this theorem is due to I. G. Petrovsky. See I. G. Petrovsky, *Lectures on the Theory of Ordinary Differential Equations*, chapter 6.

all vectors $\mathbf{x}$ orthogonal [4]) to $\mathbf{e}$, that is, all vectors $\mathbf{x}$ for which $(\mathbf{x}, \mathbf{e}) = 0$, is invariant under A. Indeed, let $\mathbf{x} \in \mathbf{R}'$, i.e., $(\mathbf{x}, \mathbf{e}) = 0$. Then

$$(A\mathbf{x}, \mathbf{e}) = (\mathbf{x}, A^*\mathbf{e}) = (\mathbf{x}, \lambda\mathbf{e}) = 0,$$

that is, $A\mathbf{x} \in \mathbf{R}'$. This proves the invariance of $\mathbf{R}'$ under A.

We now turn to the proof of Theorem 1.

Let A be a linear transformation on an $(n + 1)$-dimensional space $\mathbf{R}$. According to our lemma there exists an n-dimensional subspace $\mathbf{R}'$ of $\mathbf{R}$, invariant under A. By the induction assumption we can choose a basis in $\mathbf{R}'$ relative to which A is in canonical form. Denote this basis by

$$\mathbf{e}_1, \mathbf{e}_2, \cdots, \mathbf{e}_p; \quad \mathbf{f}_1, \mathbf{f}_2, \cdots, \mathbf{f}_q; \quad \cdots; \quad \mathbf{h}_1, \mathbf{h}_2, \ldots, \mathbf{h}_s,$$

where $p + q + \cdots + s = n$. Considered on $\mathbf{R}'$, alone, the transformation A has relative to this basis the form

$$\begin{aligned}
A\mathbf{e}_1 &= \lambda_1\mathbf{e}_1, \\
A\mathbf{e}_2 &= \mathbf{e}_1 + \lambda_1\mathbf{e}_2, \\
&\cdots\cdots\cdots \\
A\mathbf{e}_p &= \mathbf{e}_{p-1} + \lambda_1\mathbf{e}_p, \\
A\mathbf{f}_1 &= \lambda_2\mathbf{f}_1, \\
A\mathbf{f}_2 &= \mathbf{f}_1 + \lambda_2\mathbf{f}_2, \\
&\cdots\cdots\cdots \\
A\mathbf{f}_q &= \mathbf{f}_{q-1} + \lambda_2\mathbf{f}_q, \\
&\cdot \\
&\cdot \\
&\cdot \\
A\mathbf{h}_1 &= \lambda_k\mathbf{h}_1, \\
A\mathbf{h}_2 &= \mathbf{h}_1 + \lambda_k\mathbf{h}_2, \\
&\cdots\cdots\cdots \\
A\mathbf{h}_s &= \mathbf{h}_{s-1} + \lambda_k\mathbf{h}_s.
\end{aligned}$$

We now pick a vector $\mathbf{e}$ which together with the vectors

$$\mathbf{e}_1, \mathbf{e}_2, \cdots, \mathbf{e}_p; \quad \mathbf{f}_1, \mathbf{f}_2, \cdots, \mathbf{f}_q; \quad \cdots; \quad \mathbf{h}_1, \mathbf{h}_2, \cdots, \mathbf{h}_s$$

forms a basis in $\mathbf{R}$.

Applying the transformation A to $\mathbf{e}$ we get

[4] We assume here that $\mathbf{R}$ is Euclidean, i.e., that an inner product is defined on $\mathbf{R}$. However, by changing the proof slightly we can show that the Lemma holds for any vector space $\mathbf{R}$.

$$\mathrm{A}\mathbf{e} = \alpha_1 \mathbf{e}_1 + \cdots + \alpha_p \mathbf{e}_p + \beta_1 \mathbf{f}_1 + \cdots + \beta_q \mathbf{f}_q + \cdots + \delta_1 \mathbf{h}_1 + \cdots + \delta_s \mathbf{h}_s + \tau \mathbf{e}. \text{ [5]}$$

We can assume that $\tau = 0$. Indeed, if relative to some basis A is in canonical form then relative to the same basis $\mathrm{A} - \tau\mathrm{E}$ is also in canonical form and conversely. Hence if $\tau \neq 0$ we can consider the transformation $\mathrm{A} - \tau\mathrm{E}$ instead of A.

This justifies our putting

$$(1) \quad \mathrm{A}\mathbf{e} = \alpha_1 \mathbf{e}_1 + \cdots + \alpha_p \mathbf{e}_p + \beta_1 \mathbf{f}_1 + \cdots + \beta_q \mathbf{f}_q + \cdots + \delta_1 \mathbf{h}_1 + \cdots + \delta_s \mathbf{h}_s.$$

We shall now try to replace the vector $\mathbf{e}$ by some vector $\mathbf{e}'$ so that the expression for $\mathrm{A}\mathbf{e}'$ is as simple as possible. We shall seek $\mathbf{e}'$ in the form

$$(2) \quad \mathbf{e}' = \mathbf{e} - \chi_1 \mathbf{e}_1 - \cdots - \chi_p \mathbf{e}_p - \mu_1 \mathbf{f}_1 - \cdots - \mu_q \mathbf{f}_q - \cdots - \omega_1 \mathbf{h}_1 - \cdots - \omega_s \mathbf{h}_s.$$

We have

$$\mathrm{A}\mathbf{e}' = \mathrm{A}\mathbf{e} - \mathrm{A}(\chi_1 \mathbf{e}_1 + \cdots + \chi_p \mathbf{e}_p) - \mathrm{A}(\mu_1 \mathbf{f}_1 + \cdots + \mu_q \mathbf{f}_q] - \cdots - \mathrm{A}(\omega_1 \mathbf{h}_1 + \cdots + \omega_s \mathbf{h}_s),$$

or, making use of (1)

$$(3) \quad \begin{aligned} \mathrm{A}\mathbf{e}' = {} & \alpha_1 \mathbf{e}_1 + \cdots + \alpha_p \mathbf{e}_p + \beta_1 \mathbf{f}_1 + \cdots + \beta_q \mathbf{f}_q + \cdots + \delta_1 \mathbf{h}_1 \\ & + \cdots + \delta_s \mathbf{h}_s - \mathrm{A}(\chi_1 \mathbf{e}_1 + \cdots + \chi_p \mathbf{e}_p) - \mathrm{A}(\mu_1 \mathbf{f}_1 + \cdots \\ & + \mu_q \mathbf{f}_q) - \cdots - \mathrm{A}(\omega_1 \mathbf{h}_1 + \cdots + \omega_s \mathbf{h}_s). \end{aligned}$$

The coefficients $\chi_1, \cdots, \chi_p;\ \mu_1, \cdots, \mu_q;\ \cdots;\ \omega_1, \cdots, \omega_s$ can be chosen arbitrarily. We will choose them so that the right side of (3) has as few terms as possible.

We know that to each set of basis vectors in the n-dimensional space $\mathbf{R}'$ relative to which A is in canonical form there corresponds

[5] The linear transformation A has in the $(n+1)$-dimensional space $\mathbf{R}$ the eigenvalues $\lambda_1, \lambda_2, \cdots, \lambda_k$ and τ. Indeed, the matrix of A relative to the basis $\mathbf{e}_1, \mathbf{e}_2, \cdots, \mathbf{e}_p;\ \mathbf{f}_1, \mathbf{f}_2, \cdots, \mathbf{f}_q;\ \cdots;\ \mathbf{h}_1, \mathbf{h}_2, \cdots, \mathbf{h}_s, \mathbf{e}$ is triangular with the numbers $\lambda_1, \lambda_2, \cdots, \lambda_k, \tau$ on the principal diagonal.

Since the eigenvalues of a triangular matrix are equal to the entries on the diagonal (cf. for instance, § 10, para. 4) it follows that $\lambda_1, \lambda_2, \cdots, \lambda_k$, and τ are the eigenvalues of A considered on the $(n+1)$-dimensional space $\mathbf{R}$. Thus, as a result of the transition from the n-dimensional invariant subspace $\mathbf{R}'$ to the $(n+1)$-dimensional space $\mathbf{R}$ the number of eigenvalues is increased by one, namely, by the eigenvalue τ.

one eigenvalue. These eigenvalues may or may not be all different from zero. We consider first the case when all the eigenvalues are different from zero. We shall show that in this case we can choose a vector $\mathbf{e}'$ so that $A\mathbf{e}' = \mathbf{0}$, i.e., we can choose $\chi_1, \cdots, \omega_s$ so that the right side of (3) becomes zero. Assume this to be feasible. Then since the transformation A takes the vectors of each set into a linear combination of vectors of the same set it must be possible to select $\chi_1, \cdots, \omega_s$ so that the linear combination of *each set* of vectors vanishes. We show how to choose the coefficients $\chi_1, \chi_2, \cdots, \chi_p$ so that the linear combination of the vectors $\mathbf{e}_1, \cdots, \mathbf{e}_p$ in (3) vanishes. The terms containing the vectors $\mathbf{e}_1, \mathbf{e}_2, \cdots, \mathbf{e}_p$ are of the form

$$\begin{aligned}
&\alpha_1\mathbf{e}_1 + \cdots + \alpha_p\mathbf{e}_p - A(\chi_1\mathbf{e}_1 + \cdots + \chi_p\mathbf{e}_p) \\
&= \alpha_1\mathbf{e}_1 + \cdots + \alpha_p\mathbf{e}_p - \chi_1\lambda_1\mathbf{e}_1 \\
&- \chi_2(\mathbf{e}_1 + \lambda_1\mathbf{e}_2) - \cdots - \chi_p(\mathbf{e}_{p-1} + \lambda_1\mathbf{e}_p) \\
&= (\alpha_1 - \chi_1\lambda_1 - \chi_2)\mathbf{e}_1 + (\alpha_2 - \chi_2\lambda_1 - \chi_3)\mathbf{e}_2 \\
&+ \cdots + (\alpha_{p-1} - \chi_{p-1}\lambda_1 - \chi_p)\mathbf{e}_{p-1} + (\alpha_p - \chi_p\lambda_1)\mathbf{e}_p.
\end{aligned}$$

We put the coefficient of $\mathbf{e}_p$ equal to zero and determine χ_p (this can be done since $\lambda_1 \neq 0$); next we put the coefficient of $\mathbf{e}_{p-1}$ equal to zero and determine χ_{p-1}, etc. In this way the linear combination of the vectors $\mathbf{e}_1, \cdots, \mathbf{e}_p$ in (3) vanishes. The coefficients of the other sets of vectors are computed analogously.

We have thus determined $\mathbf{e}'$ so that

$$A\mathbf{e}' = \mathbf{0}.$$

By adding this vector to the basis vectors of $\mathbf{R}'$ we obtain a basis $\mathbf{e}'$; $\mathbf{e}_1, \mathbf{e}_2, \cdots, \mathbf{e}_p$; $\mathbf{f}_1, \mathbf{f}_2, \cdots, \mathbf{f}_q$; $\cdots$; $\mathbf{h}_1, \mathbf{h}_2, \cdots, \mathbf{h}_s$ in the $(n+1)$-dimensional space $\mathbf{R}$ relative to which the transformation is in canonical form. The vector $\mathbf{e}'$ forms a separate set. The eigenvalue associated with $\mathbf{e}'$ is zero (or τ if we consider the transformation A rather than $A - \tau E$).

Consider now the case when some of the eigenvalues of the transformation A on $\mathbf{R}'$ are zero. In this case the summands on the right side of (3) are of two types: those corresponding to sets of vectors associated with an eigenvalue different from zero and those associated with an eigenvalue equal to zero. The sets of the former type can be dealt with as above; i.e., for such sets we can choose

coefficients so that the appropriate linear combinations of vectors in each set vanish. Let us assume that we are left with, say, three sets of vectors,
$\mathbf{e}_1, \mathbf{e}_2, \cdots, \mathbf{e}_p$; $\mathbf{f}_1, \mathbf{f}_2, \cdots, \mathbf{f}_q$; $\mathbf{g}_1, \mathbf{g}_2, \cdots \mathbf{g}_r$ whose eigenvalues are equal to zero, i.e., $\lambda_1 = \lambda_2 = \lambda_3 = 0$. Then

$$
\begin{aligned}
(4)\qquad \mathrm{A}\mathbf{e}' = \alpha_1 \mathbf{e}_1 + \cdots + \alpha_p \mathbf{e}_p + \beta_1 \mathbf{f}_1 + \cdots + \beta_q \mathbf{f}_q + \gamma_1 \mathbf{g}_1 \\
+ \cdots + \gamma_r \mathbf{g}_r - \mathrm{A}(\chi_1 \mathbf{e}_1 + \cdots + \chi_p \mathbf{e}_p) \\
- \mathrm{A}(\mu_1 \mathbf{f}_1 + \cdots + \mu_q \mathbf{f}_q) - \mathrm{A}(\gamma_1 \mathbf{g}_1 + \cdots + \gamma_r \mathbf{g}_r).
\end{aligned}
$$

Since $\lambda_1 = \lambda_2 = \lambda_3 = 0$, it follows that

$$
\begin{array}{llll}
\mathrm{A}\mathbf{e}_1 = \mathbf{0}, & \mathrm{A}\mathbf{e}_2 = \mathbf{e}_1, & \cdots, & \mathrm{A}\mathbf{e}_p = \mathbf{e}_{p-1}, \\
\mathrm{A}\mathbf{f}_1 = \mathbf{0}, & \mathrm{A}\mathbf{f}_2 = \mathbf{f}_1, & \cdots, & \mathrm{A}\mathbf{f}_q = \mathbf{f}_{q-1}, \\
\mathrm{A}\mathbf{g}_1 = \mathbf{0}, & \mathrm{A}\mathbf{g}_2 = \mathbf{g}_1, & \cdots, & \mathrm{A}\mathbf{g}_r = \mathbf{g}_{r-1}.
\end{array}
$$

Therefore the linear combination of the vectors $\mathbf{e}_1, \mathbf{e}_2, \cdots, \mathbf{e}_p$ appearing on the right side of (4) will be of the form

$$\alpha_1 \mathbf{e}_1 + \alpha_2 \mathbf{e}_2 + \cdots + \alpha_p \mathbf{e}_p - \chi_2 \mathbf{e}_1 - \chi_3 \mathbf{e}_2 - \cdots - \chi_p \mathbf{e}_{p-1}.$$

By putting $\chi_2 = \alpha_1$, $\chi_3 = \alpha_2, \cdots, \chi_p = \alpha_{p-1}$ we annihilate all vectors except $\alpha_p \mathbf{e}_p$. Proceeding in the same manner with the sets $\mathbf{f}_1, \cdots, \mathbf{f}_q$ and $\mathbf{g}_1, \cdots, \mathbf{g}_r$ we obtain a vector $\mathbf{e}'$ such that

$$\mathrm{A}\mathbf{e}' = \alpha_p \mathbf{e}_p + \beta_q \mathbf{f}_q + \gamma_r \mathbf{g}_r.$$

It might happen that $\alpha_p = \beta_q = \gamma_r = 0$. In this case we arrive at a vector $\mathbf{e}'$ such that

$$\mathrm{A}\mathbf{e}' = \mathbf{0}$$

and just as in the first case, the transformation A is already in canonical form relative to the basis $\mathbf{e}'$; $\mathbf{e}_1, \cdots, \mathbf{e}_p$; $\mathbf{f}_1, \cdots, \mathbf{f}_q$; $\cdots$; $\mathbf{h}_1, \cdots, \mathbf{h}_s$. The vector $\mathbf{e}'$, forms a separate set and is associated with the eigenvalue zero.

Assume now that at least one of the coefficients $\alpha_p, \beta_q, \gamma_r$ is different from zero. Then, in distinction to the previous cases, it becomes necessary to change some of the basis vectors of $\mathbf{R}'$. We illustrate the procedure by considering the case $\alpha_p, \beta_q, \gamma_r \neq 0$ and $p > q > r$. We form a new set of vectors by putting $\mathbf{e}'_{p+1} = \mathbf{e}'$, $\mathbf{e}'_p = \mathrm{A}\mathbf{e}'_{p+1}, \mathbf{e}'_{p-1} = \mathrm{A}\mathbf{e}'_p, \cdots, \mathbf{e}'_1 = \mathrm{A}\mathbf{e}'_2$. Thus

$$\begin{aligned}
\mathbf{e}'_{p+1} &= \mathbf{e}', \\
\mathbf{e}'_p &= A\mathbf{e}'_{p+1} = \alpha_p \mathbf{e}_p + \beta_q \mathbf{f}_q + \gamma_r \mathbf{g}_r, \\
&\cdots\cdots\cdots\cdots\cdots\cdots\cdots\cdots \\
\mathbf{e}'_{p-r+1} &= A\mathbf{e}'_{p-r+2} = \alpha_p \mathbf{e}_{p-r+1} + \beta_q \mathbf{f}_{q-r+1} + \gamma_r \mathbf{g}_1, \\
\mathbf{e}'_{p-r} &= A\mathbf{e}'_{p-r+1} = \alpha_p \mathbf{e}_{p-r} + \beta_q \mathbf{f}_{q-r}, \\
&\cdots\cdots\cdots\cdots\cdots\cdots\cdots\cdots \\
\mathbf{e}'_1 &= A\mathbf{e}'_2 = \alpha_p \mathbf{e}_1.
\end{aligned}$$

We now replace the basis vectors $\mathbf{e}', \mathbf{e}_1, \mathbf{e}_2, \cdots, \mathbf{e}_p$ by the vectors

$$\mathbf{e}'_1, \mathbf{e}'_2, \cdots, \mathbf{e}'_p, \mathbf{e}'_{p+1}$$

and leave the other basis vectors unchanged. Relative to the new basis the transformation A is in canonical form. Note that the order of the first box has been increased by one. This completes the proof of the theorem.

While constructing the canonical form of A we had to distinguish two cases:

1. The case when the additional eigenvalue τ (we assumed $\tau = 0$) did not coincide with any of the eigenvalues $\lambda_1, \cdots, \lambda_k$. In this case a separate box of order 1 was added.

2. The case when τ coincided with one of the eigenvalues $\lambda_1, \cdots, \lambda_k$. Then it was necessary, in general, to increase the order of one of the boxes by one. If $\alpha_p = \beta_q = \gamma_r = 0$, then just as in the first case, we added a new box.

§ 20. *Elementary divisors*

In this section we shall describe a method for finding the Jordan canonical form of a transformation. The results of this section will also imply the (as yet unproved) uniqueness of the canonical form.

DEFINITION 1. *The matrices $\mathscr{A}$ and $\mathscr{A}_1 = \mathscr{C}^{-1}\mathscr{A}\mathscr{C}$, where $\mathscr{C}$ is an arbitrary non-singular matrix are said to be similar.*

If the matrix $\mathscr{A}_1$ is similar to the matrix $\mathscr{A}_2$, then $\mathscr{A}_2$ is also similar to $\mathscr{A}_1$. Indeed, let

$$\mathscr{A}_1 = \mathscr{C}^{-1}\mathscr{A}_2\mathscr{C}.$$

Then

$$\mathscr{A}_2 = \mathscr{C}\mathscr{A}_1\mathscr{C}^{-1}.$$

If we put $\mathscr{C}^{-1} = \mathscr{C}_1$, we obtain

$$\mathscr{A}_2 = \mathscr{C}_1^{-1}\mathscr{A}_1\mathscr{C}_1,$$

i.e., $\mathscr{A}_2$ is similar to $\mathscr{A}_1$.

It is easy to see that if two matrices $\mathscr{A}_1$ and $\mathscr{A}_2$ are similar to some matrix $\mathscr{A}$, then $\mathscr{A}_1$ is similar to $\mathscr{A}_2$. Indeed let

$$\mathscr{A} = \mathscr{C}_1^{-1}\mathscr{A}_1\mathscr{C}_1, \qquad \mathscr{A} = \mathscr{C}_2^{-1}\mathscr{A}_2\mathscr{C}_2.$$

Then $\mathscr{C}_1^{-1}\mathscr{A}_1\mathscr{C}_1 = \mathscr{C}_2^{-1}\mathscr{A}_2\mathscr{C}_2$, i.e.,

$$\mathscr{A}_1 = \mathscr{C}_1\mathscr{C}_2^{-1}\mathscr{A}_2\mathscr{C}_2\mathscr{C}_1^{-1}.$$

Putting $\mathscr{C}_2\mathscr{C}_1^{-1} = \mathscr{C}$, we get

$$\mathscr{A}_1 = \mathscr{C}^{-1}\mathscr{A}_2\mathscr{C},$$

i.e., $\mathscr{A}_1$ is similar to $\mathscr{A}_2$.

Let $\mathscr{A}$ be the matrix of a transformation A relative to some basis. If $\mathscr{C}$ is the matrix of transition from this basis to a new basis (§ 9), then $\mathscr{C}^{-1}\mathscr{A}\mathscr{C}$ is the matrix which represents A relative to the new basis. Thus similar matrices represent the same linear transformation relative to different bases.

We now wish to obtain *invariants* of a transformation from its matrix, i.e., expressions depending on the transformation alone. In other words, we wish to construct functions of the elements of a matrix which assume the same values for similar matrices.

One such invariant was found in § 10 where we showed that the characteristic polynomial of a matrix $\mathscr{A}$, i.e., the determinant of the matrix $\mathscr{A} - \lambda\mathscr{E}$,

$$D_n(\lambda) = |\mathscr{A} - \lambda\mathscr{E}|,$$

is the same for $\mathscr{A}$ and for any matrix similar to $\mathscr{A}$. We now construct a whole system of invariants which will include the characteristic polynomial. This will be a *complete system of invariants* in the sense that if the invariants in question are the same for two matrices then the matrices are similar.

Let $\mathscr{A}$ be a matrix of order n. The kth order minors of the matrix $\mathscr{A} - \lambda\mathscr{E}$ are certain polynomials in λ. *We denote by $D_k(\lambda)$ the greatest common divisor of those minors.*[6] We also put

[6] The greatest common divisor is determined to within a numerical multiplier. We choose $D_k(\lambda)$ to be a monic polynomial. In particular, if the kth order minors are pairwise coprime we take $D_k(\lambda)$ to be 1.

$D_0(\lambda) = 1$. In particular, $D_n(\lambda)$ is the determinant of the matrix $\mathscr{A} - \lambda\mathscr{E}$. In the sequel we show that all the $D_k(\lambda)$ are invariants.

We observe that $D_{n-1}(\lambda)$ divides $D_n(\lambda)$. Indeed, the definition of $D_{n-1}(\lambda)$ implies that all minors of order $n - 1$ are divisible by $D_{n-1}(\lambda)$. If we expand the determinant $D_n(\lambda)$ by the elements of any row we obtain a sum each of whose summands is a product of an element of the row in question by its cofactor. It follows that $D_n(\lambda)$ is indeed divisible by $D_{n-1}(\lambda)$. Similarly, $D_{n-1}(\lambda)$ is divisible by $D_{n-2}(\lambda)$, etc.

EXERCISE. Find $D_k(\lambda)$ $(k = 1, 2, 3)$ for the matrix

$$\begin{bmatrix} \lambda_0 & 1 & 0 \\ 0 & \lambda_0 & 1 \\ 0 & 0 & \lambda_0 \end{bmatrix}.$$

Answer: $D_3(\lambda) = (\lambda - \lambda_0)^3$, $D_2(\lambda) = D_1(\lambda) = 1$.

LEMMA 1. *If $\mathscr{C}$ is an arbitrary non-singular matrix then the greatest common divisors of the kth order minors of the matrices $\mathscr{A} - \lambda\mathscr{E}$, $\mathscr{C}(\mathscr{A} - \lambda\mathscr{E})$ and $(\mathscr{A} - \lambda\mathscr{E})\mathscr{C}$ are the same.*

Proof: Consider the pair of matrices $\mathscr{A} - \lambda\mathscr{E}$ and $(\mathscr{A} - \lambda\mathscr{E})\mathscr{C}$. If a_{ik} are the entries of $\mathscr{A} - \lambda\mathscr{E}$ and a'_{ik} are the entries of $(\mathscr{A} - \lambda\mathscr{E})\mathscr{C}$, then

$$a'_{ik} = \sum_{j=1}^{n} a_{ij} c_{jk},$$

i.e., the entries of any row of $(\mathscr{A} - \lambda\mathscr{E})\mathscr{C}$ are linear combinations of the rows of $\mathscr{A} - \lambda\mathscr{E}$ with coefficients from $\mathscr{C}$, i.e., independent of λ. It follows that every minor of $(\mathscr{A} - \lambda\mathscr{E})\mathscr{C}$ is the sum of minors of $\mathscr{A} - \lambda\mathscr{E}$ each multiplied by some number. Hence every divisor of the kth order minors of $\mathscr{A} - \lambda\mathscr{E}$ must divide every kth order minor of $(\mathscr{A} - \lambda\mathscr{E})\mathscr{C}$. To prove the converse we apply the same reasoning to the pair of matrices $(\mathscr{A} - \lambda\mathscr{E})\mathscr{C}$ and $[(\mathscr{A} - \lambda\mathscr{E})\mathscr{C}]\mathscr{C}^{-1} = \mathscr{A} - \lambda\mathscr{E}$. This proves that the greatest common divisors of the kth order minors of $\mathscr{A} - \lambda\mathscr{E}$ and $(\mathscr{A} - \lambda\mathscr{E})\mathscr{C}$ are the same.

LEMMA 2. *For similar matrices the polynomials $D_k(\lambda)$ are identical.*

Proof: Let $\mathscr{A}$ and $\mathscr{A}' = \mathscr{C}^{-1}\mathscr{A}\mathscr{C}$ be two similar matrices. By Lemma 1 the greatest common divisor of the kth order minors $\mathscr{A} - \lambda\mathscr{E}$ is the same as the corresponding greatest common divisor for $(\mathscr{A} - \lambda\mathscr{E})\mathscr{C}$. An analogous statement holds for the matrices $\mathscr{C}^{-1}(\mathscr{A} - \lambda\mathscr{E})$ and $\mathscr{C}^{-1}(\mathscr{A} - \lambda\mathscr{E})\mathscr{C} = \mathscr{A}' - \lambda\mathscr{E}$. Hence the $D_k(\lambda)$ for $\mathscr{A}$ and $\mathscr{A}'$ are identical.

In view of the fact that the matrices which represent a transformation in different bases are similar, we conclude on the basis of Lemma 2 that

THEOREM 1. *Let* A *be a linear transformation. Then the greatest common divisor* $D_k(\lambda)$ *of the kth order minors of the matrix* $\mathscr{A} - \lambda\mathscr{E}$, *where* $\mathscr{A}$ *represents the transformation* A *in some basis, does not depend on the choice of basis.*

We now compute the polynomials $D_k(\lambda)$ for a given linear transformation A. Theorem 1 tells us that in computing the $D_k(\lambda)$ we may use the matrix which represents A relative to an arbitrarily selected basis. We shall find it convenient to choose the basis relative to which the matrix of the transformation is in Jordan canonical form. Our task is then to compute the polynomial $D_k(\lambda)$ for the matrix $\mathscr{A}$ in Jordan canonical form.

We first find the $D_k(\lambda)$ for an nth order matrix of the form

$$\begin{bmatrix} \lambda_0 & 1 & 0 & \cdots & 0 \\ 0 & \lambda_0 & 1 & \cdots & 0 \\ \cdot & \cdot & \cdot & \cdot & \cdot \\ 0 & 0 & 0 & \cdots & 1 \\ 0 & 0 & 0 & \cdots & \lambda_0 \end{bmatrix}, \tag{1}$$

i.e., for one "box" of the canonical form. Clearly $D_n(\lambda) = (\lambda - \lambda_0)^n$. If we cross out in (1) the first column and the last row we obtain a matrix $\mathscr{A}_1$ with ones on the principal diagonal and zeros above it. Hence $D_{n-1}(\lambda) = 1$. If we cross out in $\mathscr{A}_1$ like numbered rows and columns we find that $D_{n-2}(\lambda) = \cdots = D_1(\lambda) = 1$. Thus *for an individual "box" [matrix* (1)] *the* $D_k(\lambda)$ *are*

$$(\lambda - \lambda_0)^n, 1, 1, \cdots, 1.$$

We observe further that *if* $\mathscr{B}$ *is a matrix of the form*

$$\begin{matrix} \mathscr{B}_1 & 0 \\ 0 & \mathscr{B}_2 \end{matrix},$$

where $\mathscr{B}_1$ *and* $\mathscr{B}_2$ *are of order* n_1 *and* n_2, *then the mth order non-zero minors of the matrix* $\mathscr{B}$ *are of the form*

$$\Delta_m = \Delta_{m_1}^{(1)} \Delta_{m_2}^{(2)}, \qquad m_1 + m_2 = m.$$

Here $\Delta_{m_1}^{(1)}$ *are the minors of* $\mathscr{B}_1$ *of order* m_1 *and* $\Delta_{m_2}^{(2)}$ *the minors of* $\mathscr{B}_2$ *of order* m_2. [7] Indeed, if one singles out those of the first n_1 rows which enter into the minor in question and expands it by these rows (using the theorem of Laplace), the result is zero or is of the form $\Delta_{m_1}^{(1)}\Delta_{m_2}^{(2)}$.

We shall now find the polynomials $D_k(\lambda)$ for an arbitrary matrix $\mathscr{A}$ which is in Jordan canonical form. We assume that $\mathscr{A}$ has p boxes corresponding to the eigenvalue λ_1, q boxes corresponding to the eigenvalue λ_2, etc. We denote the orders of the boxes corresponding to the eigenvalue λ_1 by $n_1, n_2, \cdots, n_p$ $(n_1 \geqq n_2 \geqq \cdots \geqq n_p)$.

Let $\mathscr{B}_i$ denote the ith box in $\mathscr{B} = \mathscr{A} - \lambda\mathscr{E}$. Then $\mathscr{B}_1$, say, is of the form

$$\mathscr{B}_1 = \begin{bmatrix} \lambda_1 - \lambda & 1 & 0 & \cdots & 0 \\ 0 & \lambda_1 - \lambda & 1 & \cdots & 0 \\ \cdots & \cdots & \cdots & \cdots & \cdots \\ 0 & 0 & 0 & \cdots & 1 \\ 0 & 0 & 0 & \cdots & \lambda_1 - \lambda \end{bmatrix}.$$

We first compute $D_n(\lambda)$, i.e., the determinant of $\mathscr{B}$. This determinant is the product of the determinants of the $\mathscr{B}_i$, i.e.,

$$D_n(\lambda) = (\lambda - \lambda_1)^{n_1+n_2+\cdots+n_p} (\lambda - \lambda_2)^{m_1+m_2+\cdots+m_q} \cdots$$

We now compute $D_{n-1}(\lambda)$. Since $D_{n-1}(\lambda)$ is a factor of $D_n(\lambda)$, it must be a product of the factors $\lambda - \lambda_1, \lambda - \lambda_2, \cdots$. The problem now is to compute the degrees of these factors. Specifically, we compute the degree of $\lambda - \lambda_1$ in $D_{n-1}(\lambda)$. We observe that any non-zero minor of $\mathscr{B} = \mathscr{A} - \lambda\mathscr{E}$ is of the form

$$\Delta_{n-1} = \Delta_{t_1}^{(1)} \Delta_{t_2}^{(2)} \cdots \Delta_{t_k}^{(k)},$$

where $t_1 + t_2 + \cdots + t_k = n - 1$ and $\Delta_{t_i}^{(i)}$ denotes the t_ith order minors of the matrix $\mathscr{B}_i$. Since the sum of the orders of the minors

[7] Of course, a non-zero kth order minor of $\mathscr{B}$ may have the form $\Delta_k{}^{(1)}$, i.e., it may be entirely made up of elements of $\mathscr{B}_1$. In this case we shall write it formally as $\Delta_k = \Delta_k{}^{(1)}\Delta_0{}^{(2)}$, where $\Delta_0{}^{(2)} = 1$.

$\Delta_{t_1}^{(1)}, \Delta_{t_2}^{(2)}, \cdots, \Delta_{t_k}^{(k)}$ is $n - 1$, exactly one of these minors is of order one lower than the order of the corresponding matrix $\mathscr{B}_i$, i.e., it is obtained by crossing out a row and a column in a box of the matrix $\mathscr{B}$. As we saw (cf. page **145**) crossing out an appropriate row and column in a box may yield a minor equal to one. Therefore it is possible to select Δ_{n-1} so that some $\Delta_{t_i}^{(i)}$ is one and the remaining minors are equal to the determinants of the appropriate boxes. It follows that in order to obtain a minor of lowest possible degree in $\lambda - \lambda_1$ it suffices to cross out a suitable row and column in the box of maximal order corresponding to λ_1. This is the box of order n_1. Thus the greatest common divisor $D_{n-1}(\lambda)$ of minors of order $n - 1$ contains $\lambda - \lambda_1$ raised to the power $n_2 + n_3 + \cdots + n_p$.

Likewise, to obtain a minor Δ_{n-2} of order $n - 2$ with lowest possible power of $\lambda - \lambda_1$ it suffices to cross out an appropriate row and column in the boxes of order n_1 and n_2 corresponding to λ_1. Thus $D_{n-2}(\lambda)$ contains $\lambda - \lambda_1$ to the power $n_3 + n_4 + \cdots + n_p$, etc. The polynomials $D_{n-p}(\lambda), D_{n-p-1}(\lambda), \cdots, D_1(\lambda)$ do not contain $\lambda - \lambda_1$ at all.

Similar arguments apply in the determination of the degrees of $\lambda - \lambda_2, \lambda - \lambda_3, \cdots$ in $D_k(\lambda)$.

We have thus proved the following result.

If the Jordan canonical form of the matrix of a linear transformation A *contains* p *boxes of order* $n_1, n_2, \cdots, n_p (n_1 \geqq n_2 \geqq \cdots \geqq n_p)$ *corresponding to the eigenvalue* λ_1, q *boxes of order* $m_1, m_2, \cdots, m_q$ $(m_1 \geqq m_2 \geqq \cdots \geqq m_q)$ *corresponding to the eigenvalue* λ_2, *etc., then*

$$D_n(\lambda) = (\lambda - \lambda_1)^{n_1+n_2+n_3+\cdots+n_p} (\lambda - \lambda_2)^{m_1+m_2+m_3+\cdots+m_q} \cdots$$
$$D_{n-1}(\lambda) = (\lambda - \lambda_1)^{n_2+n_3+\cdots+n_p} (\lambda - \lambda_2)^{m_2+m_3+\cdots+m_q} \cdots$$
$$D_{n-2}(\lambda) = (\lambda - \lambda_1)^{n_3+\cdots+n_p} (\lambda - \lambda_2)^{m_3+\cdots+m_q} \cdots$$
$$\cdots\cdots\cdots\cdots\cdots\cdots\cdots\cdots\cdots\cdots\cdots\cdots$$

Beginning with $D_{n-p}(\lambda)$ the factor $(\lambda - \lambda_1)^{\cdots}$ is replaced by one. Beginning with $D_{n-q}(\lambda)$ the factor $(\lambda - \lambda_2)^{\cdots}$ is replaced by one, etc.

In the important special case when there is exactly one box of order n_1 corresponding to the eigenvalue λ_1, exactly one box of order m_1 corresponding to the eigenvalue λ_2, exactly one box of order k_1 corresponding to the eigenvalue λ_3, etc., the $D_i(\lambda)$ have the following form:

$$\begin{aligned}
D_n(\lambda) &= (\lambda - \lambda_1)^{n_1}(\lambda - \lambda_2)^{m_1}(\lambda - \lambda_3)^{k_1} \cdots \\
D_{n-1}(\lambda) &= 1 \\
D_{n-2}(\lambda) &= 1 \\
&\cdots\cdots\cdots\cdots\cdots\cdots
\end{aligned}$$

The expressions for the $D_k(\lambda)$ show that in place of the $D_k(\lambda)$ it is more convenient to consider their ratios

$$E_k(\lambda) = \frac{D_k(\lambda)}{D_{k-1}(\lambda)}.$$

The $E_k(\lambda)$ are called *elementary divisors.* Thus *if the Jordan canonical form of a matrix $\mathscr{A}$ contains p boxes of order $n_1, n_2, \cdots, n_p (n_1 \geqq n_2 \geqq \cdots \geqq n_p)$ corresponding to the eigenvalue λ_1, q boxes of order $m_1, m_2, \cdots, m_q$ $(m_1 \geqq m_2 \geqq \cdots \geqq m_q)$ corresponding to the eigenvalue λ_2, etc., then the elementary divisors $E_k(\lambda)$ are*

$$\begin{aligned}
E_n(\lambda) &= (\lambda - \lambda_1)^{n_1} (\lambda - \lambda_2)^{m_1} \cdots, \\
E_{n-1}(\lambda) &= (\lambda - \lambda_1)^{n_2} (\lambda - \lambda_2)^{m_2} \cdots, \\
E_{n-2}(\lambda) &= (\lambda - \lambda_1)^{n_3} (\lambda - \lambda_2)^{m_3} \cdots, \\
&\cdots\cdots\cdots\cdots\cdots\cdots
\end{aligned}$$

Prescribing the elementary divisors $E_n(\lambda)$, $E_{n-1}(\lambda)$, $\cdots$, determines the Jordan canonical form of the matrix $\mathscr{A}$ uniquely. The eigenvalues λ_i are the roots of the equation $E_n(\lambda)$. The orders $n_1, n_2, \cdots, n_p$ of the boxes corresponding to the eigenvalue λ_1 coincide with the powers of $(\lambda - \lambda_1)$ in $E_n(\lambda)$, $E_{n-1}(\lambda)$, $\cdots$.

We can now state necessary and sufficient conditions for the existence of a basis in which the matrix of a linear transformation is diagonal.

A necessary and sufficient condition for the existence of a basis in which the matrix of a transformation is diagonal is that the elementary divisors have simple roots only.

Indeed, we saw that the multiplicities of the roots $\lambda_1, \lambda_2, \cdots$, of the elementary divisors determine the order of the boxes in the Jordan canonical form. Thus the simplicity of the roots of the elementary divisors signifies that all the boxes are of order one, i.e., that the Jordan canonical form of the matrix is diagonal.

THEOREM 2. *For two matrices to be similar it is necessary and sufficient that they have the same elementary divisors.*

Proof: We showed (Lemma 2) that similar matrices have the same polynomials $D_k(\lambda)$ and therefore the same elementary divisors $E_k(\lambda)$ (since the latter are quotients of the $D_k(\lambda)$).

Conversely, let two matrices $\mathscr{A}$ and $\mathscr{B}$ have the same elementary divisors. $\mathscr{A}$ and $\mathscr{B}$ are similar to Jordan canonical matrices. Since the elementary divisors of $\mathscr{A}$ and $\mathscr{B}$ are the same, their Jordan canonical forms must also be the same. This means that $\mathscr{A}$ and $\mathscr{B}$ are similar to the same matrix. But this means that $\mathscr{A}$ and $\mathscr{B}$ are similar matrices.

THEOREM 3. *The Jordan canonical form of a linear transformation is uniquely determined by the linear transformation.*

Proof: The matrices of A relative to different bases are similar. Since similar matrices have the same elementary divisors and these determine uniquely the Jordan canonical form of a matrix, our theorem follows.

We are now in a position to find the Jordan canonical form of a matrix of a linear transformation. For this it suffices to find the elementary divisors of the matrix of the transformation relative to some basis. When these are represented as products of the form $(\lambda - \lambda_1)^n(\lambda - \lambda_2)^m \cdots$ we have the eigenvalues as well as the order of the boxes corresponding to each eigenvalue.

§ 21. Polynomial matrices

1. By a polynomial matrix we mean a matrix whose entries are polynomials in some letter λ. By the degree of a polynomial matrix we mean the maximal degree of its entries. It is clear that a polynomial matrix of degree n can be written in the form

$$\mathrm{A}_0\lambda^n + \mathrm{A}_1\lambda^{n-1} + \cdots + \mathrm{A}_0,$$

where the A_k are constant matrices.[8] The matrices $\mathrm{A} - \lambda\mathrm{E}$ which we considered on a number of occasions are of this type. The results to be derived in this section contain as special cases many of the results obtained in the preceding sections for matrices of the form $\mathrm{A} - \lambda\mathrm{E}$.

[8] In this section matrices are denoted by printed Latin capitals.

Polynomial matrices occur in many areas of mathematics. Thus, for example, in solving a system of first order homogeneous linear differential equations with constant coefficients

$$\frac{dy_i}{dx} = \sum_{k=1}^{n} a_{ik} y_k \qquad (i = 1, 2, \cdots, n) \tag{1}$$

we seek solutions of the form

$$y_k = c_k e^{\lambda x}, \tag{2}$$

where λ and c_k are constants. To determine these constants we substitute the functions in (2) in the equations (1) and divide by $e^{\lambda x}$. We are thus led to the following system of linear equations:

$$\lambda c_i = \sum_{k=1}^{n} a_{ik} c_k .$$

The matrix of this system of equations is $A - \lambda E$, with A the matrix of coefficients in the system (1). Thus the study of the system of differential equations (1) is closely linked to polynomial matrices of degree one, namely, those of the form $A - \lambda E$.

Similarly, the study of higher order systems of differential equations leads to polynomial matrices of degree higher than one. Thus the study of the system

$$\sum_{k=1}^{n} a_{ik} \frac{d^2 y_k}{dx^2} + \sum_{k=1}^{n} b_{ik} \frac{dy_k}{dx} + \sum_{k=1}^{n} c_{ik} y_k = 0$$

is synonymous with the study of the polynomial matrix $A\lambda^2 + B\lambda + C$, where $A = ||a_{ik}||$, $B = ||b_{ik}||$, $C = ||c_{ik}||$.

We now consider the problem of the canonical form of polynomial matrices with respect to so-called elementary transformations.

The term "elementary" applies to the following classes of transformations.

1. Permutation of two rows or columns.
2. Addition to some row of another row multiplied by some polynomial $\varphi(\lambda)$ and, similarly, addition to some column of another column multiplied by some polynomial.
3. Multiplication of some row or column by a non-zero constant.

DEFINITION 1. *Two polynomial matrices are called equivalent if it is possible to obtain one from the other by a finite number of elementary transformations.*

The inverse of an elementary transformation is again an elementary transformation. This is easily seen for each of the three types of elementary transformations. Thus, e.g., if the polynomial matrix $B(\lambda)$ is obtained from the polynomial matrix $A(\lambda)$ by a permutation of rows then the inverse permutation takes $B(\lambda)$ into $A(\lambda)$. Again, if $B(\lambda)$ is obtained from $A(\lambda)$ by adding the ith row multiplied by $\varphi(\lambda)$ to the kth row, then $A(\lambda)$ can be obtained from $B(\lambda)$ by adding to the kth row of $B(\lambda)$ the ith row multiplied by $-\varphi(\lambda)$.

The above remark implies that if a polynomial matrix $K(\lambda)$ is equivalent to $L(\lambda)$, then $L(\lambda)$ is equivalent to $K(\lambda)$. Indeed, if $L(\lambda)$ is the result of applying a sequence of elementary transformations to $K(\lambda)$, then by applying the inverse transformations in reverse order to $L(\lambda)$ we obtain $K(\lambda)$.

If two polynomial matrices $K_1(\lambda)$ and $K_2(\lambda)$ are equivalent to a third matrix $K(\lambda)$, then they must be equivalent to each other. Indeed, by applying to $K_1(\lambda)$ first the transformations which take it into $K(\lambda)$ and then the elementary transformations which take $K(\lambda)$ into $K_2(\lambda)$, we will have taken $K_1(\lambda)$ into $K_2(\lambda)$. Thus $K_1(\lambda)$ and $K_2(\lambda)$ are indeed equivalent.

The main result of para. 1 of this section asserts the possibility of diagonalizing a polynomial matrix by means of elementary transformations. We precede the proof of this result with the following lemma:

LEMMA. *If the element $a_{11}(\lambda)$ of a polynomial matrix $A(\lambda)$ is not zero and if not all the elements $a_{ik}(\lambda)$ of $A(\lambda)$ are divisible by $a_{11}(\lambda)$, then it is possible to find a polynomial matrix $B(\lambda)$ equivalent to $A(\lambda)$ and such that $b_{11}(\lambda)$ is also different from zero and its degree is less than that of $a_{11}(\lambda)$.*

Proof: Assume that the element of $A(\lambda)$ which is not divisible by $a_{11}(\lambda)$ is in the first row. Thus let $a_{1k}(\lambda)$ not be divisible by $a_{11}(\lambda)$. Then $a_{1k}(\lambda)$ is of the form

$$a_{1k}(\lambda) = a_{11}(\lambda)\varphi(\lambda) + b(\lambda),$$

where $b(\lambda) \neq 0$ and of degree less than $a_{11}(\lambda)$. Multiplying the first column by $\varphi(\lambda)$ and subtracting the result from the kth column, we obtain a matrix with $b(\lambda)$ in place of $a_{1k}(\lambda)$, where the degree of $b(\lambda)$ is less than that of $a_{11}(\lambda)$. Permuting the first and kth columns of the new matrix puts $b(\lambda)$ in the upper left corner and results in a matrix with the desired properties. We can proceed in an analogous manner if the element not divisible by $a_{11}(\lambda)$ is in the first column.

Now let all the elements of the first row and column be divisible by $a_{11}(\lambda)$ and let $a_{ik}(\lambda)$ be an element not divisible by $a_{11}(\lambda)$. We will reduce this case to the one just considered. Since $a_{i1}(\lambda)$ is divisible by $a_{11}(\lambda)$, it must be of the form $a_{i1}(\lambda) = \varphi(\lambda)a_{11}(\lambda)$. If we subtract from the ith row the first row multiplied by $\varphi(\lambda)$, then $a_{i1}(\lambda)$ is replaced by zero and $a_{ik}(\lambda)$ is replaced by $a'_{ik}(\lambda) = a_{ik}(\lambda) - \varphi(\lambda)a_{1k}(\lambda)$ which again is not divisible by $a_{11}(\lambda)$ (this because we assumed that $a_{1k}(\lambda)$ *is* divisible by $a_{11}(\lambda)$). We now add the ith row to the first row. This leaves $a_{11}(\lambda)$ unchanged and replaces $a_{1k}(\lambda)$ with $a_{1k}(\lambda) + a'_{ik}(\lambda) = a_{1k}(\lambda)(1 - \varphi(\lambda)) + a_{ik}(\lambda)$. Thus the first row now contains an element not divisible by $a_{11}(\lambda)$ and this is the case dealt with before. This completes the proof of our lemma.

In the sequel we shall make use of the following observation. If all the elements of a polynomial matrix $\mathrm{B}(\lambda)$ are divisible by some polynomial $\mathrm{E}(\lambda)$, then all the entries of a matrix equivalent to $\mathrm{B}(\lambda)$ are again divisible by $\mathrm{E}(\lambda)$.

We are now in a position to reduce a polynomial matrix to diagonal form.

We may assume that $a_{11}(\lambda) \neq 0$. Otherwise suitable permutation of rows and columns puts a non-zero element in place of $a_{11}(\lambda)$. If not all the elements of our matrix are divisible by $a_{11}(\lambda)$, then, in view of our lemma, we can replace our matrix with an equivalent one in which the element in the upper left corner is of lower degree than $a_{11}(\lambda)$ and still different from zero. Repeating this procedure a finite number of times we obtain a matrix $\mathrm{B}(\lambda)$ all of whose elements are divisible by $b_{11}(\lambda)$.

Since $b_{12}(\lambda), \cdots, b_{1n}(\lambda)$ are divisible by $b_{11}(\lambda)$, we can, by subtracting from the second, third, etc. columns suitable multiples of the first column replace the second, third, $\cdots$, nth element of the first row with zero. Similarly, the second, third, $\cdots$, nth element of the first column can be replaced with zero. The new matrix inherits from $\mathrm{B}(\lambda)$ the property that all its entries are divisible by $b_{11}(\lambda)$. Dividing the first row by the leading coefficient of $b_{11}(\lambda)$ replaces $b_{11}(\lambda)$ with a monic polynomial $E_1(\lambda)$ but does not affect the zeros in that row.

We now have a matrix of the form

$$(3)\qquad \begin{bmatrix} E_1(\lambda) & 0 & 0 & \cdots & 0 \\ 0 & c_{22}(\lambda) & c_{23}(\lambda) & \cdots & c_{2n}(\lambda) \\ 0 & c_{32}(\lambda) & c_{33}(\lambda) & \cdots & c_{3n}(\lambda) \\ \cdots & \cdots & \cdots & \cdots & \cdots \\ 0 & c_{n2}(\lambda) & c_{n3}(\lambda) & \cdots & c_{nn}(\lambda) \end{bmatrix}$$

all of whose elements are divisible by $E_1(\lambda)$.

We can apply to the matrix $||c_{ik}||$ of order $n - 1$ the same procedure which we just applied to the matrix of order n. Then $c_{22}(\lambda)$ is replaced by a monic polynomial $E_2(\lambda)$ and the other $c_{ik}(\lambda)$ in the first row and first column are replaced with zeros. Since the entries of the larger matrix other than $E_1(\lambda)$ are zeros, an elementary transformation of the matrix of the c_{ik} can be viewed as an elementary transformation of the larger matrix. Thus we obtain a matrix whose "off-diagonal" elements in the first two rows and columns are zero and whose first two diagonal elements are monic polynomials $E_1(\lambda)$, $E_2(\lambda)$, with $E_2(\lambda)$ a multiple of $E_1(\lambda)$. Repetition of this process obviously leads to a diagonal matrix. This proves

THEOREM 1. *Every polynomial matrix can be reduced by elementary transformations to the diagonal form*

$$(4)\qquad \begin{bmatrix} E_1(\lambda) & 0 & 0 & \cdots & 0 \\ 0 & E_2(\lambda) & 0 & \cdots & 0 \\ 0 & 0 & E_3(\lambda) & \cdots & 0 \\ \cdots & \cdots & \cdots & \cdots & \cdots \\ 0 & 0 & 0 & \cdots & E_n(\lambda) \end{bmatrix}.$$

Here the diagonal elements $E_k(\lambda)$ are monic polynomials and $E_1(\lambda)$ divides $E_2(\lambda)$, $E_2(\lambda)$ divides $E_3(\lambda)$, etc. This form of a polynomial matrix is called its canonical diagonal form.

It may, of course, happen that

$$E_{r+1}(\lambda) = E_{r+2}(\lambda) = \cdots = 0$$

for some value of r.

REMARK: We have brought $A(\lambda)$ to a diagonal form in which every diagonal element is divisible by its predecessor. If we dispense with the latter requirement the process of diagonalization can be considerably simplified.

Indeed, to replace the off-diagonal elements of the first row and column with zeros it is sufficient that these elements (and not all the elements of the matrix) be divisible by $a_{11}(\lambda)$. As can be seen from the proof of the lemma this requires far fewer elementary transformations than reduction to canonical diagonal form. Once the off-diagonal elements of the first row and first column are all zero we repeat the process until we reduce the matrix to diagonal form. In this way the matrix can be reduced to various diagonal forms; i.e., the diagonal form of a polynomial matrix is not uniquely determined. On the other hand we will see in the next section that the *canonical* diagonal form of a polynomial matrix is uniquely determined.

EXERCISE. Reduce the polynomial matrix

$$\begin{bmatrix} \lambda - \lambda_1 & 0 \\ 0 & \lambda - \lambda_2 \end{bmatrix}, \quad \lambda_1 \neq \lambda_2$$

to canonical diagonal form.

Answer:

$$\begin{bmatrix} 1 & 0 \\ 0 & (\lambda - \lambda_1)(\lambda - \lambda_2) \end{bmatrix}.$$

2. In this paragraph we prove that the canonical diagonal form of a given matrix is uniquely determined. To this end we shall construct a system of polynomials connected with the given polynomial matrix which are invariant under elementary transformations and which determine the canonical diagonal form completely.

Let there be given an arbitrary polynomial matrix. Let $D_k(\lambda)$ denote the greatest common divisor of all kth order minors of the given matrix. As before, it is convenient to put $D_0(\lambda) = 1$. Since $D_k(\lambda)$ is determined to within a multiplicative constant, we take its leading coefficient to be one. In particular, if the greatest common divisor of the kth order minors is a constant, we take $D_k(\lambda) = 1$.

We shall prove that the polynomials $D_k(\lambda)$ are invariant under elementary transformations, i.e., that equivalent matrices have the same polynomials $D_k(\lambda)$.

In the case of elementary transformations of type 1 which permute rows or columns this is obvious, since such transformations either do not affect a particular kth order minor at all, or change its sign or replace it with another kth order minor. In all these cases the greatest common divisor of all kth order minors remains unchanged. Likewise, elementary transformations of type 3 do not change $D_k(\lambda)$ since under such transformations the minors are at most multiplied by a constant. Now consider elementary transformations of type 2. Specifically, consider addition of the jth column multiplied by $\varphi(\lambda)$ to the ith column. If some particular kth order minor contains none of these columns or if it contains both of them it is not affected by the transformation in question. If it contains the ith column but not the kth column we can write it as a combination of minors each of which appears in the original matrix. Thus in this case, too, the greatest common divisor of the kth order minors remains unchanged.

If all kth order minors and, consequently, all minors of order higher than k are zero, then we put $D_k(\lambda) = D_{k+1}(\lambda) = \cdots = D_n(\lambda) = 0$. We observe that equality of the $D_k(\lambda)$ for all equivalent matrices implies that equivalent matrices have the same rank.

We compute the polynomials $D_k(\lambda)$ for a matrix in canonical form

$$\begin{bmatrix} E_1(\lambda) & 0 & \cdots & 0 \\ 0 & E_2(\lambda) & \cdots & 0 \\ \cdots & \cdots & \cdots & \cdots \\ 0 & 0 & \cdots & E_n(\lambda) \end{bmatrix}. \tag{5}$$

We observe that in the case of a diagonal matrix the only non-zero minors are the principal minors, that is, minors made up of like numbered rows and columns. These minors are of the form $E_{i_1}(\lambda)E_{i_2}(\lambda) \cdots E_{i_k}(\lambda)$.

Since $E_2(\lambda)$ is divisible by $E_1(\lambda)$, $E_3(\lambda)$ is divisible by $E_2(\lambda)$, etc., it follows that the greatest common divisor $D_1(\lambda)$ of all minors of order one is $E_1(\lambda)$. Since all the polynomials $E_k(\lambda)$ are divisible by $E_1(\lambda)$ and all polynomials other than $E_1(\lambda)$ are divisible by $E_2(\lambda)$, the product $E_i(\lambda)E_j(\lambda)\,(i < j)$ is always divisible by the minor $E_1(\lambda)E_2(\lambda)$. Hence $D_2(\lambda) = E_1(\lambda)E_2(\lambda)$. Since all $E_k(\lambda)$ other than $E_1(\lambda)$ and $E_2(\lambda)$ are divisible by $E_3(\lambda)$, the product $E_i(\lambda)E_j(\lambda)E_k(\lambda) \quad (i < j < k)$ is divisible by the minor $E_1(\lambda)E_2(\lambda)E_3(\lambda)$ and so $D_3(\lambda) = E_1(\lambda)E_2(\lambda)E_3(\lambda)$.

Thus for the matrix (4)

$$(6) \qquad D_k(\lambda) = E_1(\lambda)E_2(\lambda)\cdots E_k(\lambda) \qquad (k = 1, 2, \cdots, n).$$

Clearly, if beginning with some value of r

$$E_{r+1}(\lambda) = E_{r+2}(\lambda) = \cdots = E_n(\lambda) = 0,$$

then

$$D_{r+1}(\lambda) = D_{r+2}(\lambda) = \cdots = D_n(\lambda) = 0.$$

Thus the diagonal entries of a polynomial matrix in canonical diagonal form (5) are given by the quotients

$$E_k(\lambda) = \frac{D_k(\lambda)}{D_{k-1}(\lambda)}.$$

Here, if $D_{r+1}(\lambda) = \cdots = D_n(\lambda) = 0$ we must put $E_{r+1}(\lambda) = \cdots = E_n(\lambda) = 0$.

The polynomials $E_k(\lambda)$ are called elementary divisors. In § 20 we defined the elementary divisors of matrices of the form $\mathrm{A} - \lambda\mathrm{E}$.

THEOREM 2. *The canonical diagonal form of a polynomial matrix $\mathrm{A}(\lambda)$ is uniquely determined by this matrix. If $D_k(\lambda) \neq 0$ ($k = 1, 2, \cdots, r$) is the greatest common divisor of all kth order minors of $\mathrm{A}(\lambda)$ and $D_{r+1}(\lambda) = \cdots = D_n(\lambda) = 0$, then the elements of the canonical diagonal form* (5) *are defined by the formulas*

$$E_k(\lambda) = \frac{D_k(\lambda)}{D_{k-1}(\lambda)} \qquad (k = 1, 2, \cdots, r),$$

$$E_{r+1}(\lambda) = E_{r+2}(\lambda) = \cdots = E_n(\lambda) = 0.$$

Proof: We showed that the polynomials $D_k(\lambda)$ are invariant under elementary transformations. Hence if the matrix $\mathrm{A}(\lambda)$ is equivalent to a diagonal matrix (5), then both have the same $D_k(\lambda)$. Since in the case of the matrix (5) we found that

$$D_k(\lambda) = E_1(\lambda)\cdots E_k(\lambda) \qquad (k = 1, 2, \cdots, r, r \leqq n)$$

and that

$$D_{r+1}(\lambda) = D_{r+2}(\lambda) = \cdots = D_n(\lambda) = 0,$$

the theorem follows.

COROLLARY. *A necessary and sufficient condition for two polyno-*

mial matrices $A(\lambda)$ *and* $B(\lambda)$ *to be equivalent is that the polynomials* $D_1(\lambda)$, $D_2(\lambda)$, $\cdots$, $D_n(\lambda)$ *be the same for both matrices.*

Indeed, if the polynomials $D_k(\lambda)$ are the same for $A(\lambda)$ and $B(\lambda)$, then both of these matrices are equivalent to the same canonical diagonal matrix and are therefore equivalent (to one another).

3. A polynomial matrix $P(\lambda)$ is said to be invertible if the matrix $[P(\lambda)]^{-1}$ is also a polynomial matrix. If $\det P(\lambda)$ is a constant other than zero, then $P(\lambda)$ is invertible. Indeed, the elements of the inverse matrix are, apart from sign, the $(n-1)$st order minors divided by $\det P(\lambda)$. In our case these quotients would be polynomials and $[P(\lambda)]^{-1}$ would be a polynomial matrix. Conversely, if $P(\lambda)$ is invertible, then $\det P(\lambda) = \text{const} \neq 0$. Indeed, let $[P(\lambda)]^{-1} = P_1(\lambda)$. Then $\det P(\lambda) \cdot \det P_1(\lambda) = 1$ and a product of two polynomials equals one only if the polynomials in question are non-zero constants. We have thus shown that *a polynomial matrix is invertible if and only if its determinant is a non-zero constant.*

All invertible matrices are equivalent to the unit matrix. Indeed, the determinant of an invertible matrix is a non-zero constant, so that $D_n(\lambda) = 1$. Since $D_n(\lambda)$ is divisible by $D_k(\lambda)$, $D_k(\lambda) = 1$ $(k = 1, 2, \cdots, n)$. It follows that all the elementary divisors $E_k(\lambda)$ of an invertible matrix are equal to one and the canonical diagonal form of such a matrix is therefore the unit matrix.

THEOREM 3. *Two polynomial matrices* $A(\lambda)$ *and* $B(\lambda)$ *are equivalent if and only if there exist invertible polynomial matrices* $P(\lambda)$ *and* $Q(\lambda)$ *such that.*

$$A(\lambda) = P(\lambda)B(\lambda)Q(\lambda). \tag{7}$$

Proof: We first show that if $A(\lambda)$ and $B(\lambda)$ are equivalent, then there exist invertible matrices $P(\lambda)$ and $Q(\lambda)$ such that (7) holds. To this end we observe that every elementary transformation of a polynomial matrix $A(\lambda)$ can be realized by multiplying $A(\lambda)$ on the right or on the left by a suitable invertible polynomial matrix, namely, by the matrix of the elementary transformation in question.

We illustrate this for all three types of elementary transformations. Thus let there be given a polynomial matrix $A(\lambda)$

$$A(\lambda) = \begin{bmatrix} a_{11}(\lambda) & a_{12}(\lambda) & \cdots & a_{1n}(\lambda) \\ a_{21}(\lambda) & a_{22}(\lambda) & \cdots & a_{2n}(\lambda) \\ \cdots & \cdots & \cdots & \cdots \\ a_{n1}(\lambda) & a_{n2}(\lambda) & \cdots & a_{nn}(\lambda) \end{bmatrix}.$$

To permute the first two columns (rows) of this matrix, we must multiply it on the right (left) by the matrix

$$\begin{bmatrix} 0 & 1 & 0 & \cdots & 0 \\ 1 & 0 & 0 & \cdots & 0 \\ 0 & 0 & 1 & \cdots & 0 \\ \cdots & \cdots & \cdots & \cdots & \cdots \\ 0 & 0 & 0 & \cdots & 1 \end{bmatrix} \tag{8}$$

obtained by permuting the first two columns (or, what amounts to the same thing, rows) of the unit matrix.

To multiply the second column (row) of the matrix $A(\lambda)$ by some number α we must multiply it on the right (left) by the matrix

$$\begin{bmatrix} 1 & 0 & 0 & \cdots & 0 \\ 0 & \alpha & 0 & \cdots & 0 \\ 0 & 0 & 1 & \cdots & 0 \\ \cdots & \cdots & \cdots & \cdots & \cdots \\ 0 & 0 & 0 & \cdots & 1 \end{bmatrix} \tag{9}$$

obtained from the unit matrix by multiplying its second column (or, what amounts to the same thing, row) by α.

Finally, to add to the first column of $A(\lambda)$ the second column multiplied by $\varphi(\lambda)$ we must multiply $A(\lambda)$ on the right by the matrix

$$\begin{bmatrix} 1 & 0 & 0 & \cdots & 0 \\ \varphi(\lambda) & 1 & 0 & \cdots & 0 \\ 0 & 0 & 1 & \cdots & 0 \\ \cdots & \cdots & \cdots & \cdots & \cdots \\ 0 & 0 & 0 & \cdots & 1 \end{bmatrix} \tag{10}$$

obtained from the unit matrix by just such a process. Likewise to add to the first row of $A(\lambda)$ the second row multiplied by $\varphi(\lambda)$ we must multiply $A(\lambda)$ on the left by the matrix

$$\begin{bmatrix} 1 & \varphi(\lambda) & 0 & \cdots & 0 \\ 0 & 1 & 0 & \cdots & 0 \\ 0 & 0 & 1 & \cdots & 0 \\ \cdots & \cdots & \cdots & \cdots & \cdots \\ 0 & 0 & 0 & \cdots & 1 \end{bmatrix} \tag{11}$$

obtained from the unit matrix by just such an elementary transformation.

As we see the matrices of elementary transformations are obtained by applying an elementary transformation to E. To effect an elementary transformation of the columns of a polynomial matrix $A(\lambda)$ we must multiply it by the matrix of the transformation on the right and to effect an elementary transformation of the rows of $A(\lambda)$ we must multiply it by the appropriate matrix on the left.

Computation of the determinants of the matrices (8) through (11) shows that they are all non-zero constants and the matrices are therefore invertible. Since the determinant of a product of matrices equals the product of the determinants, it follows that the product of matrices of elementary transformations is an invertible matrix.

Since we assumed that $A(\lambda)$ and $B(\lambda)$ are equivalent, it must be possible to obtain $A(\lambda)$ by applying a sequence of elementary transformations to $B(\lambda)$. Every elementary transformation can be effected by multiplying $B(\lambda)$ by an invertible polynomial matrix. Consequently, $A(\lambda)$ can be obtained from $B(\lambda)$ by multiplying the latter by some sequence of invertible polynomial matrices on the left and by some sequence of invertible polynomial matrices on the right. Since the product of invertible matrices is an invertible matrix, the first part of our theorem is proved.

It follows that every invertible matrix is the product of matrices of elementary transformations. Indeed, every invertible matrix $Q(\lambda)$ is equivalent to the unit matrix and can therefore be written in the form

$$Q(\lambda) = P_1(\lambda) E P_2(\lambda)$$

where $P_1(\lambda)$ and $P_2(\lambda)$ are products of matrices of elementary transformations. But this means that $Q(\lambda) = P_1(\lambda)P_2(\lambda)$ is itself a product of matrices of elementary transformations.

This observation can be used to prove the second half of our theorem. Indeed, let

$$A(\lambda) = P(\lambda)B(\lambda)Q(\lambda),$$

where $P(\lambda)$ and $Q(\lambda)$ are invertible matrices. But then, in view of our observation, $A(\lambda)$ is obtained from $B(\lambda)$ by applying to the latter a sequence of elementary transformations. Hence $A(\lambda)$ is equivalent to $B(\lambda)$, which is what we wished to prove.

4.[9] In this paragraph we shall study polynomial matrices of the form $A - \lambda E$, A constant. The main problem solved here is that of the equivalence of polynomial matrices $A - \lambda E$ and $B - \lambda E$ of degree one.[10]

It is easy to see that if A and B are similar, i.e., if there exists a non-singular constant matrix C such that $B = C^{-1}AC$, then the polynomial matrices $A - \lambda E$ and $B - \lambda E$ are equivalent. Indeed, if

$$B = C^{-1}AC,$$

then

$$B - \lambda E = C^{-1}(A - \lambda E)C.$$

Since a non-singular constant matrix is a special case of an invertible polynomial matrix, Theorem 3 implies the equivalence of $A - \lambda E$ and $B - \lambda E$.

Later we show the converse of this result, namely, that the equivalence of the polynomial matrices $A - \lambda E$ and $B - \lambda E$ implies the similarity of the matrices A and B. This will yield, among others, a new proof of the fact that every matrix is similar to a matrix in Jordan canonical form.

We begin by proving the following lemma:

LEMMA. *Every polynomial matrix*

$$P(\lambda) = P_0\lambda^n + P_1\lambda^{n-1} + \cdots + P_n$$

can be divided on the left by a matrix of the form $A - \lambda E$ (A *any constant matrix*); *i.e., there exist matrices* $S(\lambda)$ *and* R (R *constant*) *such that*

$$P(\lambda) = (A - \lambda E)S(\lambda) + R.$$

The process of division involved in the proof of the lemma differs from ordinary division only in that our multiplication is non-commutative.

[9] This paragraph may be omitted since it contains an alternate proof, independent of § 19, of the fact that every matrix can be reduced to Jordan canonical form.

[10] Every polynomial matrix $A_0 + \lambda A_1$ with $\det A_1 \neq 0$ is equivalent to a matrix of the form $A - \lambda E$. Indeed, in this case $A_0 + \lambda A_1 = -A_1 \times (-A_1^{-1}A_0 - \lambda E)$ and if we denote $A_1^{-1}A_0$ by A we have $A_0 + \lambda A_1 = -A_1(A - \lambda E)$ which implies (Theorem 3) the equivalence of $A_0 + \lambda A_1$ and $A - \lambda E$.

Let

$$P(\lambda) = P_0\lambda^n + P_1\lambda^{n-1} + \cdots + P_n,$$

where the P_k are constant matrices.

It is easy to see that the polynomial matrix

$$P(\lambda) + (A - \lambda E)P_0\lambda^{n-1}$$

is of degree not higher than $n - 1$.

If

$$P(\lambda) + (A - \lambda E)P_0\lambda^{n-1} = P'_0\lambda^{n-1} + P'_1\lambda^{n-2} + \cdots + P'_{n-1},$$

then the polynomial matrix

$$P(\lambda) + (A - \lambda E)P_0\lambda^{n-1} + (A - \lambda E)P'_0\lambda^{n-2}$$

is of degree not higher than $n - 2$. Continuing this process we obtain a polynomial matrix

$$P(\lambda) + (A - \lambda E)(P_0\lambda^{n-1} + P'_0\lambda^{n-2} + \cdots)$$

of degree not higher than zero, i.e., independent of λ. If R denotes the constant matrix just obtained, then

$$P(\lambda) = (A - \lambda E)(-P_0\lambda^{n-1} - P'_0\lambda^{n-2} + \cdots) + R,$$

or putting $S(\lambda) = (-P_0\lambda^{n-1} - P'_0\lambda^{n-2} + \cdots)$,

$$P(\lambda) = (A - \lambda E)S(\lambda) + R.$$

This proves our lemma.

A similar proof holds for the possibility of division on the right; i.e., there exist matrices $S_1(\lambda)$ and R_1 such that

$$P(\lambda) = S_1(\lambda)(A - \lambda E) + R_1.$$

We note that in our case, just as in the ordinary theorem of Bezout, we can claim that

$$R = R_1 = P(A).$$

THEOREM 4. *The polynomial matrices* $A - \lambda E$ *and* $B - \lambda E$ *are equivalent if and only if the matrices* A *and* B *are similar.*

Proof: The sufficiency part of the proof was given in the beginning of this paragraph. It remains to prove necessity. This means that we must show that the equivalence of $A - \lambda E$ and $B - \lambda E$ implies the similarity of A and B. By Theorem 3 there exist invertible polynomial matrices $P(\lambda)$ and $Q(\lambda)$ such that

$$\text{(12)} \qquad B - \lambda E = P(\lambda)(A - \lambda E)Q(\lambda).$$

We shall first show that $P(\lambda)$ and $Q(\lambda)$ in (12) may be replaced by constant matrices.

To this end we divide $P(\lambda)$ on the left by $B - \lambda E$ and $Q(\lambda)$ by $B - \lambda E$ on the right. Then

$$\text{(13)} \qquad \begin{aligned} P(\lambda) &= (B - \lambda E)P_1(\lambda) + P_0, \\ Q(\lambda) &= Q_1(\lambda)(B - \lambda E) + Q_0, \end{aligned}$$

where P_0 and Q_0 are constant matrices.

If we insert these expressions for $P(\lambda)$ and $Q(\lambda)$ in the formula (12) and carry out the indicated multiplications we obtain

$$\begin{aligned} B - \lambda E = (B - \lambda E)P_1(\lambda)(A - \lambda E)Q_1(\lambda)(B - \lambda E) \\ + (B - \lambda E)P_1(\lambda)(A - \lambda E)Q_0 + P_0(A - \lambda E)Q_1(\lambda)(B - \lambda E) \\ + P_0(A - \lambda E)Q_0. \end{aligned}$$

If we transfer the last summand on the right side of the above equation to its left side and denote the sum of the remaining terms by $K(\lambda)$, i.e., if we put

$$\text{(14)} \qquad \begin{aligned} K(\lambda) = (B - \lambda E)P_1(\lambda)(A - \lambda E)Q_1(\lambda)(B - \lambda E) \\ + (B - \lambda E)P_1(\lambda)(A - \lambda E)Q_0 \\ + P_0(A - \lambda E)Q_1(\lambda)(B - \lambda E), \end{aligned}$$

then we get

$$\text{(15)} \qquad B - \lambda E - P_0(A - \lambda E)Q_0 = K(\lambda).$$

Since $Q_1(\lambda)(B - \lambda E) + Q_0 = Q(\lambda)$, the first two summands in $K(\lambda)$ can be written as follows:

$$\begin{aligned} (B - \lambda E)P_1(\lambda)(A - \lambda E)Q_1(\lambda)(B - \lambda E) \\ + (B - \lambda E)P_1(\lambda)(A - \lambda E)Q_0 = (B - \lambda E)P_1(\lambda)(A - \lambda E)Q(\lambda). \end{aligned}$$

We now add and subtract from the third summand in $K(\lambda)$ the expression $(B - \lambda E)P_1(\lambda)(A - \lambda E)Q_1(\lambda)(B - \lambda E)$ and find

$$\text{(16)} \qquad \begin{aligned} K(\lambda) = (B - \lambda E)P_1(\lambda)(A - \lambda E)Q(\lambda) \\ + P(\lambda)(A - \lambda E)Q_1(\lambda)(B - \lambda E) \\ -(B - \lambda E)P_1(\lambda)(A - \lambda E)Q_1(\lambda)(B - \lambda E). \end{aligned}$$

But in view of (12)

$$\begin{aligned} (A - \lambda E)Q(\lambda) &= P^{-1}(\lambda)(B - \lambda E), \\ P(\lambda)(A - \lambda E) &= (B - \lambda E)Q^{-1}(\lambda). \end{aligned}$$

Using these relations we can rewrite $K(\lambda)$ in the following manner

$$K(\lambda) = (B - \lambda E)[P_1(\lambda)P^{-1}(\lambda) + Q^{-1}(\lambda)Q_1(\lambda) - P_1(\lambda)(A - \lambda E)Q_1(\lambda)](B - \lambda E).$$

We now show that $K(\lambda) = 0$. Since $P(\lambda)$ and $Q(\lambda)$ are invertible, the expression in square brackets is a polynomial in λ. We shall prove this polynomial to be zero. Assume that this polynomial is not zero and is of degree m. Then it is easy to see that $K(\lambda)$ is of degree $m + 2$ and since $m \geqq 0$, $K(\lambda)$ is at least of degree two. But (15) implies that $K(\lambda)$ is at most of degree one. Hence the expression in the square brackets, and with it $K(\lambda)$, is zero.

We have thus found that

$$B - \lambda E = P_0(A - \lambda E)Q_0, \tag{17}$$

where P_0 and Q_0 are constant matrices; i.e., we may indeed replace $P(\lambda)$ and $Q(\lambda)$ in (12) with constant matrices.

Equating coefficients of λ in (17) we see that

$$P_0Q_0 = E,$$

which shows that the matrices P_0 and Q_0 are non-singular and that

$$P_0 = Q_0^{-1}.$$

Equating the free terms we find that

$$B = P_0AQ_0 = Q_0^{-1}AQ_0,$$

i.e., that A and B are similar. This completes the proof of our theorem.

Since equivalence of the matrices $A - \lambda E$ and $B - \lambda E$ is synonymous with identity of their elementary divisors it follows from the theorem just proved that *two matrices* A *and* B *are similar if and only if the matrices* $A - \lambda E$ *and* $B - \lambda E$ *have the same elementary divisors.* We now show that *every matrix* A *is similar to a matrix in Jordan canonical form.*

To this end we consider the matrix $A - \lambda E$ and find its elementary divisors. Using these we construct as in § 20 a matrix B in Jordan canonical form. $B - \lambda E$ has the same elementary divisors as $A - \lambda E$, but then B is similar to A.

As was indicated on page 160 (footnote) this paragraph gives another proof of the fact that every matrix is similar to a matrix in Jordan canonical form. Of course, the contents of this paragraph can be deduced directly from §§ 19 and 20.

CHAPTER IV

Introduction to Tensors

§ 22. *The dual space*

1. *Definition of the dual space.* Let **R** be a vector space. Together with **R** one frequently considers another space called the dual space which is closely connected with **R**. The starting point for the definition of a dual space is the notion of a linear function introduced in para. 1, § 4.

We recall that a function $f(\mathbf{x})$, $\mathbf{x} \in \mathbf{R}$, is called linear if it satisfies the following conditions:

1. $f(\mathbf{x} + \mathbf{y}) = f(\mathbf{x}) + f(\mathbf{y})$,
2. $f(\lambda \mathbf{x}) = \lambda f(\mathbf{x})$.

Let $\mathbf{e}_1, \mathbf{e}_2, \cdots, \mathbf{e}_n$ be a basis in an n-dimensional space **R**. If

$$\mathbf{x} = \xi^1 \mathbf{e}_1 + \xi^2 \mathbf{e}_2 + \cdots + \xi^n \mathbf{e}_n$$

is a vector in **R** and f is a linear function on **R**, then (cf. § 4) we can write

$$(1) \qquad \begin{aligned} f(\mathbf{x}) &= f(\xi^1 \mathbf{e}_1 + \xi^2 \mathbf{e}_2 + \cdots + \xi^n \mathbf{e}_n) \\ &= a_1 \xi^1 + a_2 \xi^2 + \cdots + a_n \xi^n, \end{aligned}$$

where the coefficients $a_1, a_2, \cdots, a_n$ which determine the linear function are given by

$$(2) \qquad a_1 = f(\mathbf{e}_1), \qquad a_2 = f(\mathbf{e}_2), \qquad \cdots, \qquad a_n = f(\mathbf{e}_n).$$

It is clear from (1) that *given a basis* $\mathbf{e}_1, \mathbf{e}_2, \cdots, \mathbf{e}_n$ *every n-tuple* $a_1, a_2, \cdots, a_n$ *determines a unique linear function.*

Let f and g be linear functions. By the sum h of f and g we mean the function which associates with a vector $\mathbf{x}$ the number $f(\mathbf{x}) + g(\mathbf{x})$. By the product of f by a number α we mean the function which associates with a vector $\mathbf{x}$ the number $\alpha f(\mathbf{x})$.

Obviously the sum of two linear functions and the product of a function by a number are again linear functions. Also, if f is determined by the numbers $a_1, a_2, \cdots, a_n$ and g by the numbers $b_1, b_2, \cdots, b_n$, then $f + g$ is determined by the numbers $a_1 + b_1$, $a_2 + b_2, \cdots, a_n + b_n$ and αf by the numbers $\alpha a_1, \alpha a_2, \cdots, \alpha a_n$. Thus the totality of linear functions on **R** forms a vector space.

DEFINITION 1. *Let* **R** *be an n-dimensional vector space. By the dual space* $\overline{\mathbf{R}}$ *of* **R** *we mean the vector space whose elements are linear functions defined on* **R**. *Addition and scalar multiplication in* $\overline{\mathbf{R}}$ *follow the rules of addition and scalar multiplication for linear functions.*

In view of the fact that relative to a given basis $\mathbf{e}_1, \mathbf{e}_2, \cdots, \mathbf{e}_n$ in **R** every linear function f is uniquely determined by an n-tuple $a_1, a_2, \cdots, a_n$ and that this correspondence preserves sums and products (of vectors by scalars), it follows that $\overline{\mathbf{R}}$ is isomorphic to the space of n-tuples of numbers.

One consequence of this fact is that *the dual space* $\overline{\mathbf{R}}$ *of the n-dimensional space* **R** *is likewise n-dimensional.*

The vectors in **R** are said to be *contravariant*, those in $\overline{\mathbf{R}}$, *covariant*. In the sequel $\mathbf{x}, \mathbf{y}, \cdots$ will denote elements of **R** and $f, g, \cdots$ elements of $\overline{\mathbf{R}}$.

2. *Dual bases*. In the sequel we shall denote the value of a linear function f at a point $\mathbf{x}$ by $(f, \mathbf{x})$. Thus with every pair $f \in \overline{\mathbf{R}}$ and $\mathbf{x} \in \mathbf{R}$ there is associated a number $(f, \mathbf{x})$ and the following relations hold:

1. $(f, \mathbf{x}_1 + \mathbf{x}_2) = (f, \mathbf{x}_1) + (f, \mathbf{x}_2)$,
2. $(f, \lambda\mathbf{x}) = \lambda(f, \mathbf{x})$,
3. $(\lambda f, \mathbf{x}) = \lambda(f, \mathbf{x})$,
4. $(f_1 + f_2, \mathbf{x}) = (f_1, \mathbf{x}) + (f_2, \mathbf{x})$.

The first two of these relations stand for $f(\mathbf{x}_1+\mathbf{x}_2)=f(\mathbf{x}_1)+f(\mathbf{x}_2)$ and $f(\lambda\mathbf{x}) = \lambda f(\mathbf{x})$ and so express the linearity of f. The third defines the product of a linear function by a number and the fourth, the sum of two linear functions. The form of the relations 1 through 4 is like that of Axioms 2 and 3 for an inner product (§ 2). However, an inner product is a number associated with a pair of vectors from the same Euclidean space whereas $(f, \mathbf{x})$ is a number associated with a pair of vectors belonging to two different vector spaces $\overline{\mathbf{R}}$ and **R**.

Two vectors $\mathbf{x} \in \mathbf{R}$ and $f \in \overline{\mathbf{R}}$ are said to be *orthogonal* if

$$(f, \mathbf{x}) = 0.$$

In the case of a single space $\mathbf{R}$ orthogonality is defined for Euclidean spaces only. If $\mathbf{R}$ is an arbitrary vector space we can still speak of elements of $\mathbf{R}$ being orthogonal to elements of $\overline{\mathbf{R}}$.

DEFINITION 2. *Let* $\mathbf{e}_1, \mathbf{e}_2, \cdots, \mathbf{e}_n$ *be a basis in* $\mathbf{R}$ *and* $f^1, f^2, \cdots, f^n$ *a basis in* $\overline{\mathbf{R}}$. *The two bases are said to be dual if*

$$(3) \qquad (f^i, \mathbf{e}_k) = \begin{cases} 1 & \text{when } i = k \\ 0 & \text{when } i \neq k \end{cases} \qquad (i, k = 1, 2, \cdots, n).$$

In terms of the symbol $\delta_k^{\ i}$, defined by

$$\delta_k^{\ i} = \begin{cases} 1 \text{ when } i = k \\ 0 \text{ when } i \neq k \end{cases} \qquad (i, k = 1, 2, \cdots, n),$$

condition (3) can be rewritten as

$$(f^i, \mathbf{e}_k) = \delta_k^{\ i}.$$

If $\mathbf{e}_1, \mathbf{e}_2, \cdots, \mathbf{e}_n$ is a basis in $\mathbf{R}$, then $(f, \mathbf{e}_k) = f(\mathbf{e}_k)$ give the numbers a_k which determine the linear function $f \in \overline{\mathbf{R}}$ (cf. formula (2)). This remark implies that

if $\mathbf{e}_1, \mathbf{e}_2, \cdots, \mathbf{e}_n$ *is a basis in* $\mathbf{R}$, *then there exists a unique basis* $f^1, f^2, \cdots, f^n$ *in* $\overline{\mathbf{R}}$ *dual to* $\mathbf{e}_1, \mathbf{e}_2, \cdots, \mathbf{e}_n$.

The proof is immediate: The equations

$$(f^1, \mathbf{e}_1) = 1, \qquad (f^1, \mathbf{e}_2) = 0, \qquad \cdots, \qquad (f^1, \mathbf{e}_n) = 0$$

define a unique vector (linear function) $f^1 \in \overline{\mathbf{R}}$. The equations

$$(f^2, \mathbf{e}_1) = 0, \qquad (f^2, \mathbf{e}_2) = 1, \qquad \cdots, \qquad (f^2, \mathbf{e}_n) = 0$$

define a unique vector (linear function) $f^2 \in \overline{\mathbf{R}}$, etc. The vectors $f^1, f^2, \cdots, f^n$ are linearly independent since the corresponding n-tuples of numbers are linearly independent. Thus $f^1, f^2, \cdots, f^n$ constitute a unique basis of $\overline{\mathbf{R}}$ dual to the basis $\mathbf{e}_1, \mathbf{e}_2, \cdots, \mathbf{e}_n$ of $\mathbf{R}$.

In the sequel we shall follow a familiar convention of tensor analysis according to which one leaves out summation signs and sums over any index which appears as a superscript and a subscript. Thus $\xi^i \eta_i$ stands for $\xi^1 \eta_1 + \xi^2 \eta_2 + \cdots + \xi^n \eta_n$.

Given dual bases $\mathbf{e}_i$ and f_k one can easily express the coordinates of any vector. Thus, if $\mathbf{x} \in \mathbf{R}$ and

$$\mathbf{x} = \xi^i \mathbf{e}_i,$$

then

$$(f^k, \mathbf{x}) = (f^k, \xi^i \mathbf{e}_i) = \xi^i (f^k, \mathbf{e}_i) = \xi^i \delta_i^k = \xi^k.$$

Hence, *the coordinates* ξ^k *of a vector* $\mathbf{x}$ *in the basis* $\mathbf{e}_1, \mathbf{e}_2, \cdots, \mathbf{e}_n$ *can be computed from the formulas*

$$\xi^k = (f^k, \mathbf{x}),$$

where f^k is the basis dual to the basis $\mathbf{e}_i$.

Similarly, if $f \in \overline{\mathbf{R}}$ and

$$f = \eta_k f^k,$$

then

$$\eta_i = (f, \mathbf{e}_i).$$

Now let $\mathbf{e}_1, \mathbf{e}_2, \cdots, \mathbf{e}_n$ and $f^1, f^2, \cdots, f^n$ be dual bases. We shall express the number $(f, \mathbf{x})$ in terms of the coordinates of the vectors f and $\mathbf{x}$ with respect to the bases $\mathbf{e}_1, \mathbf{e}_2, \cdots, \mathbf{e}_n$ and $f^1, f^2, \cdots, f^n$, respectively. Thus let

$$\mathbf{x} = \xi^1 \mathbf{e}_1 + \xi^2 \mathbf{e}_2 + \cdots + \xi^n \mathbf{e}_n \quad \text{and} \quad f = \eta_1 f^1 + \eta_2 f^2 + \cdots + \eta_n f^n.$$

Then

$$(f, \mathbf{x}) = (\eta_i f^i, \xi^k \mathbf{e}_k) = (f^i, \mathbf{e}_k)\eta_i \xi^k = \delta_i^k \eta_i \xi^k = \eta_i \xi^i.$$

To repeat:

If $\mathbf{e}_1, \mathbf{e}_2, \cdots, \mathbf{e}_n$ *is a basis in* $\mathbf{R}$ *and* $f^1, f^2, \cdots, f^n$ its dual basis in $\overline{\mathbf{R}}$ then

$$(f, \mathbf{x}) = \eta_1 \xi^1 + \eta_2 \xi^2 + \cdots + \eta_n \xi^n, \tag{4}$$

where $\xi^1, \xi^2, \cdots, \xi^n$ *are the coordinates of* $\mathbf{x} \in \mathbf{R}$ *relative to the basis* $\mathbf{e}_1, \mathbf{e}_2, \cdots, \mathbf{e}_n$ *and* $\eta_1, \eta_2, \cdots, \eta_n$ *are the coordinates of* $f \in \overline{\mathbf{R}}$ *relative to the basis* $f^1, f^2, \cdots, f^n$.

NOTE. For arbitrary bases $\mathbf{e}_1, \mathbf{e}_2, \cdots, \mathbf{e}_n$ and $f^1, f^2, \cdots, f^n$ in $\mathbf{R}$ and $\overline{\mathbf{R}}$ respectively

$$(f, \mathbf{x}) = a_k^i \eta_i \xi^k,$$

where $a_k^i = (f^i, \mathbf{e}_k)$.

3. *Interchangeability of* $\mathbf{R}$ *and* $\overline{\mathbf{R}}$. We now show that it is possible to interchange the roles of $\mathbf{R}$ and $\overline{\mathbf{R}}$ without affecting the theory developed so far.

$\overline{\mathbf{R}}$ was defined as the totality of linear functions on $\mathbf{R}$. We wish to show that if φ is a linear function on $\overline{\mathbf{R}}$, then $\varphi(f) = (f, \mathbf{x}_0)$ for some fixed vector $\mathbf{x}_0$ in $\mathbf{R}$.

To this end we choose a basis $\mathbf{e}_1, \mathbf{e}_2, \cdots, \mathbf{e}_n$ in $\mathbf{R}$ and denote its dual by $f^1, f^2, \cdots, f^n$. If the coordinates of f relative to the basis $f^1, f^2, \cdots, f^n$ are $\eta_1, \eta_2, \cdots, \eta_n$, then we can write

$$\varphi(f) = a^1\eta_1 + a^2\eta_2 + \cdots + a^n\eta_n.$$

Now let $\mathbf{x}_0$ be the vector $a^1\mathbf{e}_1 + a^2\mathbf{e}_2 + \cdots + a^n\mathbf{e}_n$. Then, as we saw in para. 2,

$$(f, \mathbf{x}_0) = a^1\eta_1 + a^2\eta_2 + \cdots + a^n\eta_n$$

and

$$\varphi(f) \equiv (f, \mathbf{x}_0). \tag{5}$$

This formula establishes the desired one-to-one correspondence between the linear functions φ on $\overline{\mathbf{R}}$ and the vectors $\mathbf{x}_0 \in \mathbf{R}$ and permits us to view $\mathbf{R}$ as the space of linear functions on $\overline{\mathbf{R}}$ thus placing the two spaces on the same footing.

We observe that the only operations used in the simultaneous study of a space and its dual space are the operations of addition of vectors and multiplication of a vector by a scalar in each of the spaces involved and the operation $(f, \mathbf{x})$ which connects the elements of the two spaces. It is therefore possible to give a definition of a pair of dual spaces $\mathbf{R}$ and $\overline{\mathbf{R}}$ which emphasizes the parallel roles played by the two spaces. Such a definition runs as follows: a pair of dual spaces $\mathbf{R}$ and $\overline{\mathbf{R}}$ is a pair of n-dimensional vector spaces and an operation $(f, \mathbf{x})$ which associates with $f \in \overline{\mathbf{R}}$ and $\mathbf{x} \in \mathbf{R}$ a number $(f, \mathbf{x})$ so that conditions 1 through 4 above hold and, in addition,

5. *$(f, \mathbf{x}) = 0$ for all $\mathbf{x}$ implies $f = 0$, and $(f, \mathbf{x}) = 0$ for all f implies $\mathbf{x} = 0$.*

NOTE: In para. 2 above we showed that for every basis in $\mathbf{R}$ there exists a unique dual basis in $\overline{\mathbf{R}}$. In view of the interchangeability of $\mathbf{R}$ and $\overline{\mathbf{R}}$, for every basis in $\overline{\mathbf{R}}$ there exists a unique dual basis in $\mathbf{R}$.

4. *Transformation of coordinates in* $\mathbf{R}$ *and* $\overline{\mathbf{R}}$. If we specify the coordinates of a vector $\mathbf{x} \in \mathbf{R}$ relative to some basis $\mathbf{e}_1, \mathbf{e}_2, \cdots, \mathbf{e}_n$, then, as a rule, we specify the coordinates of a vector $f \in \overline{\mathbf{R}}$ relative to the dual basis $f^1, f^2, \cdots, f^n$ of $\mathbf{e}_1, \mathbf{e}_2, \cdots, \mathbf{e}_n$.

Now let $\mathbf{e}'_1, \mathbf{e}'_2, \cdots, \mathbf{e}'_n$ be a new basis in $\mathbf{R}$ whose connection with the basis $\mathbf{e}_1, \mathbf{e}_2, \cdots, \mathbf{e}_n$ is given by

$$\mathbf{e}'_i = c_i^k \mathbf{e}_k. \tag{6}$$

Let $f^1, f^2, \cdots, f^n$ be the dual basis of $\mathbf{e}_1, \mathbf{e}_2, \cdots, \mathbf{e}_n$ and $f'^1, f'^2, \cdots, f'^n$ be the dual basis of $\mathbf{e}'_1, \mathbf{e}'_2, \cdots, \mathbf{e}'_n$. We wish to find the matrix $||b_i^{\ k}||$ of transition from the f^i basis to the f'^i basis. We first find its inverse, the matrix $||u_i^{\ k}||$ of transition from the basis $f'^1, f'^2, \cdots, f'^n$ to the basis $f^1, f^2, \cdots, f^n$:

$$f^k = u_i^{\ k} f'^i. \tag{6'}$$

To this end we compute $(f^k, \mathbf{e}'_i)$ in two ways:

$$(f^k, \mathbf{e}'_i) = (f^k, c_i^{\ \alpha}\mathbf{e}_\alpha) = c_i^{\ \alpha}(f^k, \mathbf{e}_\alpha) = c_i^{\ k},$$
$$(f^k, \mathbf{e}'_i) = (u_j^{\ k} f'^j, \mathbf{e}'_i) = u_j^{\ k}(f'^j, \mathbf{e}'_i) = u_i^{\ k}.$$

Hence $c_i^{\ k} = u_i^{\ k}$, i.e., the matrix in (6′) is the transpose [1] of the transition matrix in (6). It follows that *the matrix of the transition*

$$f'^k = b_i^{\ k} f^i \tag{7}$$

from $f^1, f^2, \cdots, f^n$ *to* $f'^1, f'^2, \cdots, f'^n$ *is equal to the inverse of the transpose of the matrix* $||c_i^{\ k}||$ *which is the matrix of transition from* $\mathbf{e}_1, \mathbf{e}_2, \cdots, \mathbf{e}_n$ *to* $\mathbf{e}'_1, \mathbf{e}'_2, \cdots, \mathbf{e}'_n$.

We now discuss the effect of a change of basis on the coordinates of vectors in $\mathbf{R}$ and $\overline{\mathbf{R}}$. Thus let ξ^i be the coordinates of $\mathbf{x} \in \mathbf{R}$ relative to a basis $\mathbf{e}_1, \mathbf{e}_2, \cdots, \mathbf{e}_n$ and ξ'^i its coordinates in a new basis $\mathbf{e}'_1, \mathbf{e}'_2, \cdots, \mathbf{e}'_n$. Then

$$(f^i, \mathbf{x}) = (f^i, \xi^k \mathbf{e}_k) = \xi^i,$$
$$(f'^i, \mathbf{x}) = (f'^i, \xi'^k \mathbf{e}'_k) = \xi'^i.$$

Now

$$\xi'^i = (f'^i, \mathbf{x}) = (b_k^{\ i} f^k, \mathbf{x}) = b_k^{\ i}(f^k, \mathbf{x}) = b_k^{\ i}\xi^k,$$

so that

$$\xi'^i = b_k^{\ i}\xi^k. \tag{8}$$

It follows that the coordinates of vectors in $\mathbf{R}$ transform like the vectors of the dual basis in $\overline{\mathbf{R}}$. Similarly, the coordinates of vectors in $\overline{\mathbf{R}}$ transform like the vectors of the dual basis in $\mathbf{R}$, i.e.,

$$\eta'_i = c_i^{\ k}\eta_k. \tag{9}$$

[1] This is seen by comparing the matrices in (6) and (6′). We say that the matrix $||u_i^{\ k}||$ in (6′) is the transpose of the transition matrix in (6) because the summation indices in (6) and (6′) are different.

We summarize our findings in the following rule: *when we change from an "old" coordinate system to a "new" one objects with lower case index transform in one way and objects with upper case index transform in a different way. Of the matrices* $||c_i^k||$ *and* $||b_i^k||$ *involved in these transformations one is the inverse of the transpose of the other.*

The fact the $||b_i^k||$ is the inverse of the transpose of $||c_i^k||$ is expressed in the relations

$$c_i^\alpha b_\alpha^j = \delta_i^j, \qquad b_i^\alpha c_\alpha^j = \delta_i^j.$$

5. *The dual of a Euclidean space.* For the sake of simplicity we restrict our discussion to the case of real Euclidean spaces.

LEMMA. *Let* **R** *be an n-dimensional Euclidean space. Then every linear function f on* **R** *can be expressed in the form*

$$f(\mathbf{x}) = (\mathbf{x}, \mathbf{y}),$$

where **y** *is a fixed vector uniquely determined by the linear function f. Conversely, every vector* **y** *determines a linear function f such that* $f(\mathbf{x}) = (\mathbf{x}, \mathbf{y})$.

Proof: Let $\mathbf{e}_1, \mathbf{e}_2, \cdots, \mathbf{e}_n$ be an orthonormal basis of **R**. If $\mathbf{x} = \xi^i \mathbf{e}_i$, then $f(\mathbf{x})$ is of the form

$$f(\mathbf{x}) = a_1\xi^1 + a_2\xi^2 + \cdots + a_n\xi^n.$$

Now let **y** be the vector with coordinates $a_1, a_2, \cdots, a_n$. Since the basis $\mathbf{e}_1, \mathbf{e}_2, \cdots, \mathbf{e}_n$ is orthonormal,

$$(\mathbf{x}, \mathbf{y}) = a_1\xi^1 + a_2\xi^2 + \cdots + a_n\xi^n.$$

This shows the existence of a vector **y** such that for all **x**

$$f(\mathbf{x}) = (\mathbf{x}, \mathbf{y}).$$

To prove the uniqueness of **y** we observe that if

$$f(\mathbf{x}) = (\mathbf{x}, \mathbf{y}_1) \quad \text{and} \quad f(\mathbf{x}) = (\mathbf{x}, \mathbf{y}_2),$$

then $(\mathbf{x}, \mathbf{y}_1) = (\mathbf{x}, \mathbf{y}_2)$, i.e., $(\mathbf{x}, \mathbf{y}_1 - \mathbf{y}_2) = 0$ for all **x**. But this means that $\mathbf{y}_1 - \mathbf{y}_2 = \mathbf{0}$.

The converse is obvious.

Thus in the case of a Euclidean space every f in $\overline{\mathbf{R}}$ can be replaced with the appropriate **y** in **R** and instead of writing $(f, \mathbf{x})$ we can write $(\mathbf{y}, \mathbf{x})$. *Since the sumultaneous study of a vector space and its dual space involves only the usual vector operations and the operation*

$(f, \mathbf{x})$ *which connects elements* $f \in \overline{\mathbf{R}}$ *and* $\mathbf{x} \in \mathbf{R}$, *we may, in case of a Euclidean space, replace* f *by* $\mathbf{y}$, $\overline{\mathbf{R}}$ *by* $\mathbf{R}$, *and* $(f, \mathbf{x})$ *by* $(\mathbf{y}, \mathbf{x})$, *i.e., we may identify a Euclidean space* $\mathbf{R}$ *with its dual space* $\overline{\mathbf{R}}$. [2] This situation is sometimes described as follows: in Euclidean space one can replace covariant vectors by contravariant vectors.

When we identify $\mathbf{R}$ and its dual $\overline{\mathbf{R}}$ the concept of orthogonality of a vector $\mathbf{x} \in \mathbf{R}$ and a vector $f \in \overline{\mathbf{R}}$ (introduced in para. 2 above) reduces to that of orthogonality of two vectors of $\mathbf{R}$.

Let $\mathbf{e}_1, \mathbf{e}_2, \cdots, \mathbf{e}_n$ be an arbitrary basis in $\mathbf{R}$ and $f^1, f^2, \cdots, f^n$ its dual basis in $\overline{\mathbf{R}}$. If $\mathbf{R}$ is Euclidean, we can identify $\overline{\mathbf{R}}$ with $\mathbf{R}$ and so look upon the f^k as elements of $\mathbf{R}$. It is natural to try to find expressions for the f^k in terms of the given $\mathbf{e}_i$. Let

$$\mathbf{e}_i = g_{i\alpha} f^\alpha.$$

We wish to find the coefficients g_{ik}. Now

$$(\mathbf{e}_i, \mathbf{e}_k) = (g_{i\alpha} f^\alpha, \mathbf{e}_k) = g_{i\alpha}(f^\alpha, \mathbf{e}_k) = g_{i\alpha}\delta_k{}^\alpha = g_{ik}.$$

Thus if the basis of the f^i is dual to that of the $\mathbf{e}_k$, then

$$\mathbf{e}_k = g_{k\alpha} f^\alpha, \tag{10}$$

where

$$g_{ik} = (\mathbf{e}_i, \mathbf{e}_k).$$

Solving equation (10) for f^i we obtain the required result

$$f^i = g^{i\alpha}\mathbf{e}_\alpha. \tag{11}$$

where the matrix $||g^{ik}||$ is the inverse of the matrix $||g_{ik}||$, i.e.,

$$g^{i\alpha} g_{\alpha k} = \delta_k{}^i.$$

EXERCISE. Show that

$$g^{ik} = (f^i, f^k).$$

§ 23. *Tensors*

1. *Multilinear functions.* In the first chapter we studied linear and bilinear functions on an n-dimensional vector space. A natural

[2] If $\mathbf{R}$ is an n-dimensional vector space, then $\overline{\mathbf{R}}$ is also n-dimensional and so $\mathbf{R}$ and $\overline{\mathbf{R}}$ are isomorphic. If we were to identify $\mathbf{R}$ and $\overline{\mathbf{R}}$ we would have to write in place of $(f, \mathbf{x})$, $(\mathbf{y}, \mathbf{x})$, $\mathbf{y}, \mathbf{x} \in \mathbf{R}$. But this would have the effect of introducing an inner product in $\mathbf{R}$.

generalization of these concepts is the concept of a multilinear function of an arbitrary number of vectors some of which are elements of **R** and some of which are elements of $\overline{\mathbf{R}}$.

DEFINITION 1. *A function*

$$l(\mathbf{x}, \mathbf{y}, \cdots; f, g, \cdots)$$

is said to be a multilinear function of p vectors $\mathbf{x}, \mathbf{y}, \cdots \in \mathbf{R}$ and q vectors $f, g, \cdots \in \overline{\mathbf{R}}$ (the dual of **R***) if l is linear in each of its arguments.*

Thus, for example, if we fix all vectors but the first then

$$\begin{aligned} &l(\mathbf{x}' + \mathbf{x}'', \mathbf{y}, \cdots; f, g, \cdots) \\ &= l(\mathbf{x}', \mathbf{y}, \cdots; f, g, \cdots) + l(\mathbf{x}'', \mathbf{y}, \cdots; f, g, \cdots); \\ &l(\lambda\mathbf{x}, \mathbf{y}, \cdots; f, g, \cdots) = \lambda l(\mathbf{x}, \mathbf{y}, \cdots; f, g, \cdots). \end{aligned}$$

Again,

$$\begin{aligned} &l(\mathbf{x}, \mathbf{y}, \cdots; f' + f'', g, \cdots) \\ &= l(\mathbf{x}, \mathbf{y}, \cdots; f', g, \cdots) + l(\mathbf{x}, \mathbf{y}, \cdots; f'', g, \cdots); \\ &l(\mathbf{x}, \mathbf{y}, \cdots; \mu f, g, \cdots) = \mu l(\mathbf{x}, \mathbf{y}, \cdots; f, g, \cdots). \end{aligned}$$

A multilinear function of p vectors in **R** (contravariant vectors) and q vectors in $\overline{\mathbf{R}}$ (covariant vectors) is called *a multilinear function of type* (p, q).

The simplest multilinear functions are those of type (1, 0) and (0, 1).

A multilinear function of type (1, 0) is a linear function of one vector in **R**, i.e., a vector in $\overline{\mathbf{R}}$ (a covariant vector).

Similarly, as was shown in para. 3, § 22, a multilinear function of type (0, 1) defines a vector in **R** (a contravariant vector).

There are three types of multilinear functions of two vectors (bilinear functions):

(α) bilinear functions on **R** (considered in § 4),
(β) bilinear functions on $\overline{\mathbf{R}}$,
(γ) functions of one vector in **R** and one in $\overline{\mathbf{R}}$.

There is a close connection between functions of type (γ) and linear transformations. Indeed, let

$$\mathbf{y} = A\mathbf{x}$$

be a linear transformation on **R**. The bilinear function of type (γ)

associated with A is the function

$$(f, \mathbf{A}\mathbf{x})$$

which depends linearly on the vectors $\mathbf{x} \in \mathbf{R}$ and $f \in \overline{\mathbf{R}}$.

As in § 11 of chapter II one can prove the converse, i.e.,that one can associate with every bilinear function of type (γ) a linear transformation on $\mathbf{R}$.

2. *Expressions for multilinear functions in a given coordinate system. Coordinate transformations.* We now express a multilinear function in terms of the coordinates of its arguments. For simplicity we consider the case of a multilinear function $l(\mathbf{x}, \mathbf{y}; f)$, $\mathbf{x}, \mathbf{y} \in \mathbf{R}$, $f \in \overline{\mathbf{R}}$ (a function of type (2, 1)).

Let $\mathbf{e}_1, \mathbf{e}_2, \cdots, \mathbf{e}_n$ be a basis in $\mathbf{R}$ and $f^1, f^2, \cdots, f^n$ its dual in $\overline{\mathbf{R}}$. Let

$$\mathbf{x} = \xi^i \mathbf{e}_i, \qquad \mathbf{y} = \eta^j \mathbf{e}_j, \qquad f = \zeta_k f^k.$$

Then

$$l(\mathbf{x}, \mathbf{y}; f) = l(\xi^i \mathbf{e}_i, \eta^j \mathbf{e}_j; \zeta_k f^k) = \xi^i \eta^j \zeta_k l(\mathbf{e}_i, \mathbf{e}_j; f^k),$$

or

$$l(\mathbf{x}, \mathbf{y}; f) = a_{ij}{}^k \xi^i \eta^j \zeta_k,$$

where the coefficients $a_{ij}{}^k$ which determine the function $l(\mathbf{x}, \mathbf{y}; f)$ are given by the relations

$$a_{ij}{}^k = l(\mathbf{e}_i, \mathbf{e}_j; f^k).$$

This shows that the $a_{ij}{}^k$ depend on the choice of bases in $\mathbf{R}$ and $\overline{\mathbf{R}}$.

A similar formula holds for a general multilinear function

$$l(\mathbf{x}, \mathbf{y}, \cdots; f, g, \cdots) = a_{ij\cdots}^{rs\cdots} \xi^i \eta^j \cdots \lambda_r \mu_s \cdots, \tag{1}$$

where the numbers $a_{ij\cdots}^{rs\cdots}$ which define the multilinear function are given by

$$a_{ij\cdots}^{rs\cdots} = l(\mathbf{e}_i, \mathbf{e}_j, \cdots; f^r, f^s, \cdots). \tag{2}$$

We now show how the system of numbers which determine a multilinear form changes as a result of a change of basis.

Thus let $\mathbf{e}_1, \mathbf{e}_2, \cdots, \mathbf{e}_n$ be a basis in $\mathbf{R}$ and $f^1, f^2, \cdots, f^n$ its dual basis in $\overline{\mathbf{R}}$. Let $\mathbf{e}'_1, \mathbf{e}'_2, \cdots, \mathbf{e}'_n$ be a new basis in $\mathbf{R}$ and $f'^1, f'^2, \cdots, f'^n$ be its dual in $\overline{\mathbf{R}}$. If

$$\mathbf{e}'_\alpha = c_\alpha{}^\beta \mathbf{e}_\beta, \tag{3}$$

then (cf. para. 4, § 22)

$$f'^{\beta} = b_\alpha^{\beta} f^\alpha, \tag{4}$$

where the matrix $||b_\alpha^{\beta}||$ is the transpose of the inverse of $||c_\alpha^{\beta}||$.

For a fixed α the numbers c_α^{β} in (3) are the coordinates of the vector $\mathbf{e}'_\alpha$ relative to the basis $\mathbf{e}_1, \mathbf{e}_2, \cdots, \mathbf{e}_n$. Similarly, for a fixed β the numbers b_α^{β} in (4) are the coordinates of f'^{β} relative to the basis $f^1, f^2, \cdots, f^n$.

We shall now compute the numbers $a'^{rs\cdots}_{ij\cdots}$ which define our multilinear function relative to the bases $\mathbf{e}'_1, \mathbf{e}'_2, \cdots, \mathbf{e}'_n$ and $f'^1, f'^2, \cdots, f'^n$. We know that

$$a'^{rs\cdots}_{ij\cdots} = l(\mathbf{e}'_i, \mathbf{e}'_j, \cdots; f'^r, f'^s, \cdots).$$

Hence to find $a'^{rs\cdots}_{ij\cdots}$ we must put in (1) in place of $\xi^i, \eta^j, \cdots; \lambda_r, \mu_s, \cdots$ the coordinates of the vectors $\mathbf{e}'_i, \mathbf{e}'_j, \cdots; f'^r, f'^s, \cdots$, i.e., the numbers $c_i^\alpha, c_j^\beta, \cdots; b_\sigma^r, b_\tau^s, \cdots$ In this way we find that

$$a'^{rs\cdots}_{ij\cdots} = c_i^\alpha c_j^\beta \cdots b_\sigma^r b_\tau^s \cdots a^{\sigma\tau\cdots}_{\alpha\beta\cdots}.$$

To sum up: *If* $a^{rs\cdots}_{ij\cdots}$ *define a multilinear function* $l(\mathbf{x}, \mathbf{y}, \cdots; f, g, \cdots)$ *relative to a pair of dual bases* $\mathbf{e}_1, \mathbf{e}_2, \cdots \mathbf{e}_n$ *and* $f^1, f^2, \cdots, f^n$, *and* $a'^{rs\cdots}_{ij\cdots}$ *define this function relative to another pair of dual bases* $\mathbf{e}'_1, \mathbf{e}'_2, \cdots, \mathbf{e}'_n$ *and* $f'^1, f'^2, \cdots, f'^n$, *then*

$$a'^{rs\cdots}_{ij\cdots} = c_i^\alpha c_j^\beta \cdots b_\sigma^r b_\tau^s \cdots a^{\sigma\tau\cdots}_{\alpha\beta\cdots}. \tag{5}$$

Here $||c_i^j||$ *is the matrix defining the transformation of the* $\mathbf{e}$ *basis and* $||b_i^j||$ *is the matrix defining the transformation of the* f *basis.*

This situation can be described briefly by saying that *the lower indices of the numbers* $a^{rs\cdots}_{ij\cdots}$ *are affected by the matrix* $||c_i^j||$ *and the upper by the matrix* $||b_i^j||$ (cf. para. 4, § 22).

3. *Definition of a tensor.* The objects which we have studied in this book (vectors, linear functions, linear transformations, bilinear functions, etc.) were defined relative to a given basis by an appropriate system of numbers. Thus relative to a given basis a vector was defined by its n coordinates, a linear function by its n coefficients, a linear transformation by the n^2 entries in its matrix, and a bilinear function by the n^2 entries in *its* matrix. In the case of each of these objects the associated system of numbers would, upon a change of basis, transform in a manner peculiar to each object and to characterize the object one had to prescribe the

values of these numbers relative to some basis as well as their law of transformation under a change of basis.

In para. 1 and 2 of this section we introduced the concept of a multilinear function. Relative to a definite basis this object is defined by n^k numbers (2) which under change of basis transform in accordance with (5). We now define a closely related concept which plays an important role in many branches of physics, geometry, and algebra.

DEFINITION 2. *Let* **R** *be an n-dimensional vector space. We say that a p times covariant and q times contravariant tensor is defined if with every basis in* **R** *there is associated a set of* n^{p+q} *numbers* $a_{ij\cdots}^{rs\cdots}$ *(there are p lower indices and q upper indices) which under change of basis defined by some matrix* $\|c_i^j\|$ *transform according to the rule*

$$a'^{rs\cdots}_{ij\cdots} = c_i^{\alpha} c_j^{\beta} \cdots b_{\sigma}^{r} b_{\tau}^{s} \cdots a_{\alpha\beta\cdots}^{\sigma\tau\cdots} \tag{6}$$

with $\|b_i^j\|$ *the transpose of the inverse of* $\|c_i^j\|$. *The number* $p + q$ *is called the rank (valence) of the tensor. The numbers* $a_{ij\cdots}^{rs\cdots}$ *are called the components of the tensor.*

Since the system of numbers defining a multilinear function of p vectors in **R** and q vectors in $\overline{\mathbf{R}}$ transforms under change of basis in accordance with (6) the multilinear function determines a unique tensor of rank $p + q$, p times covariant and q times contravariant. Conversely, every tensor determines a unique multilinear function. This permits us to deduce properties of tensors and of the operations on tensors using the "model" supplied by multilinear functions. Clearly, multilinear functions are only one of the possible realizations of tensors.

We now give a few examples of tensors.

1. Scalar. If we associate with every coordinate system the same constant a, then a may be regarded as a tensor of rank zero. A tensor of rank zero is called a scalar.

2. Contravariant vector. Given a basis in **R** every vector in **R** determines n numbers, its coordinates relative to this basis. These transform according to the rule

$$\eta'^i = b_j^i \eta^j$$

and so represent a contravariant tensor of rank 1.

3. Linear function (covariant vector). The numbers a_i defining a linear function transform according to the rule

$$a'_i = c_i^j a_j$$

and so represent a covariant tensor of rank 1.

4. Bilinear function. Let $A(\mathbf{x}; \mathbf{y})$ be a bilinear form on **R.** With every basis we associate the matrix of the bilinear form relative to this basis. The resulting tensor is of rank two, twice covariant. Similarly, a bilinear form of vectors $\mathbf{x} \in \mathbf{R}$ and $y \in \overline{\mathbf{R}}$ defines a tensor of rank two, once covariant and once contravariant and a bilinear form of vectors $f, g \in \overline{\mathbf{R}}$ defines a twice contravariant tensor.

5. Linear transformation. Let A be a linear transformation on **R**. With every basis we associate the matrix of A relative to this basis. We shall show that this matrix is a tensor of rank two, once covariant and once contravariant.

Let $||a_i^k||$ be the matrix of A relative to some basis $\mathbf{e}_1, \mathbf{e}_2, \cdots, \mathbf{e}_n$, i.e.,

$$\mathrm{A}\mathbf{e}_i = a_i^k \mathbf{e}_k.$$

Define a change of basis by the equations

$$\mathbf{e}'_i = c_i^\alpha \mathbf{e}_\alpha.$$

Then

$$\mathbf{e}_i = b_i^\alpha \mathbf{e}'_\alpha, \qquad \text{where } b_i^\alpha c_\alpha^k = \delta_i^k.$$

It follows that

$$\mathrm{A}\mathbf{e}'_i = \mathrm{A}c_i^\alpha \mathbf{e}_\alpha = c_i^\alpha \mathrm{A}\mathbf{e}_\alpha = c_i^\alpha a_\alpha^\beta \mathbf{e}_\beta = c_i^\alpha a_\alpha^\beta b_\beta^k \mathbf{e}'_k = a'_i \mathbf{e}'_k.$$

This means that the matrix $||a'^k_i||$ of A relative to the $\mathbf{e}'_i$ basis takes the form

$$a'^k_i = a_\alpha^\beta c_i^\alpha b_\beta^k,$$

which proves that the matrix of a linear transformation is indeed a tensor of rank two, once covariant and once contravariant.

In particular, the matrix of the identity transformation E relative to any basis is the unit matrix, i.e., the system of numbers

$$\delta_i^k = \begin{cases} 1 & \text{if } i = k, \\ 0 & \text{if } i \neq k. \end{cases}$$

Thus δ_i^k is the simplest tensor of rank two once covariant and once contravariant. One interesting feature of this tensor is that its components do not depend on the choice of basis.

EXERCISE. Show dirctly that the system of numbers

$$\delta_i^k = \begin{cases} 1 & \text{if} \quad i = k, \\ 0 & \text{if} \quad i \neq k, \end{cases}$$

associated with every bais is a tensor.

We now prove two simple properties of tensors.

A sufficient condition for the equality of two tensors of the same type is the equality of their corresponding components relative to some basis. (This means that if the components of these two tensors relative to some basis are equal, then their components relative to any other basis must be equal.) For proof we observe that since the two tensors are of the same type they transform in exactly the same way and since their components are the same in some coordinate system they must be the same in every coordinate system. We wish to emphasize that the assumption about the two tensors being of the same type is essential. Thus, given a basis, both a linear transformation and a bilinear form are defined by a matrix. Coincidence of the matrices defining these objects in one basis does not imply coincidence of the matrices defining these objects in another basis.

Given p and q it is always possible to construct a tensor of type (p, q) whose components relative to some basis take on n^{p+q} prescribed values. The proof is simple. Thus let $a_{ij\cdots}^{rs\cdots}$ be the numbers prescribed in some basis. These numbers define a multilinear function $l(\mathbf{x}, \mathbf{y}, \cdots; f, g, \cdots)$ as per formula (1) in para. 2 of this section. The multilinear function, in turn, defines a unique tensor satisfying the required conditions.

4. *Tensors in Euclidean space.* If $\mathbf{R}$ is a (real) n-dimensional Euclidean space, then, as was shown in para. 5 of § 22, it is possible to establish an isomorphism between $\mathbf{R}$ and $\overline{\mathbf{R}}$ such that if $\mathbf{y} \in \mathbf{R}$ corresponds under this isomorphism to $f \in \overline{\mathbf{R}}$, then

$$(f, \mathbf{x}) = (\mathbf{y}, \mathbf{x})$$

for all $\mathbf{x} \in \mathbf{R}$. Given a multilinear function l of p vectors $\mathbf{x}, \mathbf{y}, \cdots$ in $\mathbf{R}$ and q vectors $f, g, \cdots$ in $\overline{\mathbf{R}}$ we can replace the latter by corresponding vectors $\mathbf{u}, \mathbf{v}, \cdots$ in $\mathbf{R}$ and so obtain a multilinear function $l(\mathbf{x}, \mathbf{y}, \cdots; \mathbf{u}, \mathbf{v}, \cdots)$ of $p + q$ vectors in $\mathbf{R}$.

We now propose to express the coefficients of $l(\mathbf{x}, \mathbf{y}, \cdots; \mathbf{u}, \mathbf{v}, \cdots)$ in terms of the coefficients of $l(\mathbf{x}, \mathbf{y}, \cdots; f, g, \cdots)$.

Thus let $a_{ij\cdots}^{rs\cdots}$ be the coefficients of the multilinear function $l(\mathbf{x}, \mathbf{y}, \cdots; f, g, \cdots)$, i.e.,

$$a_{ij\cdots}^{rs\cdots} = l(\mathbf{e}_i, \mathbf{e}_j, \cdots; f^r, f^s, \cdots)$$

and let $b_{rs\cdots ij\cdots}$ be the coefficients of the multilinear function $l(\mathbf{x}, \mathbf{y}, \cdots; \mathbf{u}, \mathbf{v}, \cdots)$, i.e.,

$$b_{ij\cdots rs\cdots} = l(\mathbf{e}_i, \mathbf{e}_j, \cdots; \mathbf{e}_r, \mathbf{e}_s, \cdots).$$

We showed in para. 5 of § 22 that in Euclidean space the vectors $\mathbf{e}_k$ of a basis dual to f^i are expressible in terms of the vectors f^i in the following manner:

$$\mathbf{e}_r = g_{r\alpha} f^\alpha,$$

where

$$g_{ik} = (\mathbf{e}_i, \mathbf{e}_k).$$

It follows that

$$\begin{aligned} b_{ij\cdots rs\cdots} &= l(\mathbf{e}_i, \mathbf{e}_j, \cdots; \mathbf{e}_r, \mathbf{e}_s, \cdots) \\ &= l(\mathbf{e}_i, \mathbf{e}_j, \cdots; g_{\alpha r} f^\alpha, g_{\beta s} f^\beta, \cdots) \\ &= g_{\alpha r} g_{\beta s} \cdots l(\mathbf{e}_i, \mathbf{e}_j, \cdots; f^\alpha, f^\beta, \cdots) \\ &= g_{\alpha r} g_{\beta s} \cdots a_{ij\cdots}^{\alpha\beta\cdots}. \end{aligned}$$

In view of the established connection between multilinear functions and tensors we can restate our result for tensors:

If $a_{ij\cdots}^{rs\cdots}$ is a tensor in Euclidean space p times covariant and q times contravariant, then this tensor can be used to construct a new tensor $b_{ij\cdots rs\cdots}$ which is $p + q$ times covariant. This operation is referred to as lowering of indices. It is defined by the equation

$$b_{ij\cdots rs\cdots} = g_{\alpha r} g_{\beta s}\, a_{ij\cdots}^{\alpha\beta\cdots}.$$

Here g_{ik} is a twice covariant tensor. This is obvious if we observe that the $g_{ik} = (\mathbf{e}_i, \mathbf{e}_k)$ are the coefficients of a bilinear form, namely, the inner product relative to the basis $\mathbf{e}_1, \mathbf{e}_2, \cdots, \mathbf{e}_n$. In view of its connection with the inner product (metric) in our space, *the tensor g_{ik} is called a metric tensor.*

The equation

$$b^{ij\cdots rs\cdots} = g^{\alpha i} g^{\beta j} \cdots a_{\alpha\beta\cdots}^{rs\cdots}$$

defines the analog of the operation just discussed. The new

operation is referred to as raising the indices. Here g^{ik} has the meaning discussed in para. 5 of § 22.

EXERCISE. Show that g^{ik} is a twice contravariant tensor.

5. *Operations on tensors.* In view of the connection between tensors and multilinear functions it is natural first to define operations on multilinear functions and then express these definitions in the language of tensors relative to some basis.

Addition of tensors. Let

$$l'(\mathbf{x}, \mathbf{y}, \cdots; f, g, \cdots), \qquad l''(\mathbf{x}, \mathbf{y}, \cdots; f, g, \cdots)$$

be two multilinear functions of the same number of vectors in $\mathbf{R}$ and the same number of vectors in $\overline{\mathbf{R}}$. We define their sum $l(\mathbf{x}, \mathbf{y}, \cdots; f, g, \cdots)$ by the formula

$$l(\mathbf{x}, \mathbf{y}, \cdots; f, g, \cdots) = l'(\mathbf{x}, \mathbf{y}, \cdots; f, g, \cdots) + l''(\mathbf{x}, \mathbf{y}, \cdots; f, g, \cdots).$$

Clearly this sum is again a multilinear function of the same number of vectors in $\mathbf{R}$ and $\overline{\mathbf{R}}$ as the summands l' and l''. Consequently addition of tensors is defined by means of the formula

$$a^{rs\cdots}_{ij\cdots} = a'^{rs\cdots}_{ij\cdots} + a''^{rs\cdots}_{ij\cdots}.$$

Multiplication of tensors. Let

$$l'(\mathbf{x}, \mathbf{y}, \cdots; f, g, \cdots) \quad \text{and} \quad l''(\mathbf{z}, \cdots; h, \cdots)$$

be two multilinear functions of which the first depends on p' vectors in $\mathbf{R}$ and q' vectors in $\overline{\mathbf{R}}$ and the second on p'' vectors in $\mathbf{R}$ and q'' vectors in $\overline{\mathbf{R}}$. We define the product $l(\mathbf{x}, \mathbf{y}, \cdots, \mathbf{z}, \cdots; f, g, \cdots, h, \cdots)$ of l' and l'' by means of the formula:

$$l(\mathbf{x}, \mathbf{y}, \cdots, \mathbf{z}, \cdots; f, g, \cdots, h, \cdots) = l'(\mathbf{x}, \mathbf{y}, \cdots; f, g, \cdots) l(\mathbf{z}, \cdots; h, \ldots).$$

l is a multilinear function of $p' + p''$ vectors in $\mathbf{R}$ and $q' + q''$ vectors in $\overline{\mathbf{R}}$. To see this we need only vary in l one vector at a time keeping all other vectors fixed.

We shall now express the components of the tensor corresponding to the product of the multilinear functions l' and l'' in terms of the components of the tensors corresponding to l' and l''. Since

$$a'^{rs\cdots}_{ij\cdots} = l'(\mathbf{e}_i, \mathbf{e}_j, \cdots; f^r, f^s, \cdots)$$

and

$$a''^{tu\cdots}_{kl\cdots} = l''(\mathbf{e}_k, \mathbf{e}_l, \cdots; f^t, f^u, \cdots),$$

it follows that

$$a^{rs\cdots\, tu\cdots}_{ij\cdots\, kl\cdots} = a'^{rs\cdots}_{ij\cdots} a''^{tu\cdots}_{kl\cdots}.$$

This formula defines the product of two tensors.

Contraction of tensors. Let $l(\mathbf{x}, \mathbf{y}, \cdots; f, g, \cdots)$ be a multilinear function of p vectors in $\mathbf{R}$ $(p \geqq 1)$ and q vectors in $\overline{\mathbf{R}} (q \geqq 1)$. We use l to define a new multilinear function of $p - 1$ vectors in $\mathbf{R}$ and $q - 1$ vectors in $\overline{\mathbf{R}}$. To this end we choose a basis $\mathbf{e}_1, \mathbf{e}_2, \cdots, \mathbf{e}_n$ in $\mathbf{R}$ and its dual basis $f^1, f^2, \cdots, f^n$ in $\overline{\mathbf{R}}$ and consider the sum

$$\begin{aligned} (7) \qquad & l'(\mathbf{y}, \cdots; g, \cdots) \\ &= l(\mathbf{e}_1, \mathbf{y}, \cdots; f^1, g, \cdots) + l(\mathbf{e}_2, \mathbf{y}, \cdots; f^2, g, \cdots) \\ &\quad + \cdots + l(\mathbf{e}_n, \mathbf{y}, \cdots; f^n, g, \cdots) \\ &= l(\mathbf{e}_\alpha, \mathbf{y}, \cdots; f^\alpha, g, \cdots). \end{aligned}$$

Since each summand is a multilinear function of $\mathbf{y}, \cdots$ and $g, \cdots$ the same is true of the sum l'. We now show that whereas each summand depends on the choice of basis, the sum does not. Let us choose a new basis $\mathbf{e}'_1, \mathbf{e}'_2, \cdots, \mathbf{e}'_n$ and denote its dual basis by $f'^1, f'^2, \cdots f'^n$. Since the vectors $\mathbf{y}, \cdots$ and $g, \cdots$ remain fixed we need only prove our contention for a bilinear form $A(\mathbf{x}; f)$. Specifically we must show that

$$A(\mathbf{e}_\alpha; f^\alpha) = A(\mathbf{e}'_\alpha; f'^\alpha).$$

We recall that if

$$\mathbf{e}'_i = c_i^k \mathbf{e}_k,$$

then

$$f^k = c_i^k f'^i.$$

Therefore

$$\begin{aligned} A(\mathbf{e}'_\alpha; f'^\alpha) &= A(c_\alpha^k \mathbf{e}_k; f'^\alpha) = c_\alpha^k A(\mathbf{e}_k; f'^\alpha) \\ &= A(\mathbf{e}_k; c_\alpha^k f'^\alpha) = A(\mathbf{e}_k; f^k), \end{aligned}$$

i.e., $A(\mathbf{e}_\alpha; f^\alpha)$ is indeed independent of choice of basis.

We now express the coefficients of the form (7) in terms of the coefficients of the form $l(\mathbf{x}, \mathbf{y}, \cdots; f, g, \cdots)$. Since

$$a'^{s\cdots}_{j\cdots} = l'(\mathbf{e}_j, \cdots; f^s, \cdots)$$

and

$$l'(\mathbf{e}_j, \cdots; f^s, \cdots) = l(\mathbf{e}_\alpha, \mathbf{e}_j, \cdots; f^\alpha, f^s, \cdots),$$

if follows that

$$a'^{s\cdots}_{j\cdots} = a^{\alpha s\cdots}_{\alpha j\cdots}. \tag{8}$$

The tensor $a'^{s\cdots}_{j\cdots}$ *obtained from* $a^{rs\cdots}_{ij\cdots}$ *as per* (8) *is called a contraction of the tensor* $a^{rs\cdots}_{ij\cdots}$.

It is clear that the summation in the process of contraction may involve any covariant index and any contravariant index. However, if one tried to sum over two covariant indices, say, the resulting system of numbers would no longer form a tensor (for upon change of basis this system of numbers would not transform in accordance with the prescribed law of transformation for tensors).

We observe that contraction of a tensor of rank two leads to a tensor of rank zero (scalar), i.e., to a number independent of coordinate systems.

The operation of lowering indices discussed in para. 4 of this section can be viewed as contraction of the product of some tensor by the metric tensor g_{ik} (repeated as a factor an appropriate number of times). Likewise the raising of indices can be viewed as contraction of the product of some tensor by the tensor g^{ik}.

Another example. Let $a_{ij}{}^k$ be a tensor of rank three and $b_l{}^m$ a tensor of rank two. Their product $c^{km}_{ijl} = a_{ij}{}^k b_l{}^m$ is a tensor rank five. The result of contracting this tensor over the indices i and m, say, would be a tensor of rank three. Another contraction, over the indices j and k, say, would lead to a tensor of rank one (vector).

Let $a_i{}^j$ and $b_k{}^l$ be two tensors of rank two. By multiplication and contraction these yield a new tensor of rank two:

$$c_i{}^l = a_i{}^\alpha b_\alpha{}^l.$$

If the tensors $a_i{}^j$ and $b_k{}^l$ are looked upon as matrices of linear transformations, then the tensor $c_i{}^l$ is the matrix of the product of these linear transformations.

With any tensor $a_i{}^j$ of rank two we can associate a sequence of invariants (i.e., numbers independent of choice of basis, simply scalars):

$$a_\alpha{}^\alpha,\ a_\alpha{}^\beta a_\beta{}^\alpha,\ \cdots.$$

The operations on tensors permit us to construct from given tensors new tensors invariantly connected with the given ones. For example, by multiplying vectors we can obtain tensors of arbitrarily high rank. Thus, if ξ^i are the coordinates of a contravariant vector and η_j of a covariant vector, then $\xi^i \eta_j$ is a tensor of rank two, etc. We observe that not all tensors can be obtained by multiplying vectors. However, it can be shown that every tensor can be obtained from vectors (tensors of rank one) using the operations of addition and multiplication.

By a rational integral invariant of a given system of tensors we mean a polynomial function of the components of these tensors whose value does not change when one system of components of the tensors in question computed with respect to some basis is replaced by another system computed with respect to some other basis.

In connection with the above concept we quote without proof the following result:

Any rational integral invariant of a given system of tensors can be obtained from these tensors by means of the operations of tensor multiplication, addition, multiplication by a number and total contraction (i.e., contraction over all indices).

6. *Symmetric and skew symmetric tensors*

DEFINITION. *A tensor is said to be symmetric with respect to a given set of indices*[1] *if its components are invariant under an arbitrary permutation of these indices.*

For example, if

$$a^{st\cdots}_{ik\cdots} = a^{st\cdots}_{ki\cdots}$$

then the tensor is said to be symmetric with respect to the first two (lower) indices.

If $l(\mathbf{x}, \mathbf{y}, \cdots; f, g, \cdots)$ is the multilinear function corresponding to the tensor $a^{st\cdots}_{ik\cdots}$, i.e., if

$$l(\mathbf{x}, \mathbf{y}, \cdots; f, g, \cdots) = a^{st\cdots}_{ik\cdots} \tag{9}$$

then, as is clear from (9), symmetry of the tensor with respect to some group of indices is equivalent to symmetry of the corresponding multilinear function with respect to an appropriate set of vectors. Since for a multilinear function to be symmetric with

[1] It goes without saying that we have in mind indices in the same (upper or lower) group.

respect to a certain set of vectors it is sufficient that the corresponding tensor $a_{ik\cdots}^{st\cdots}$ be symmetric with respect to an appropriate set of indices in some basis, it follows that *if the components of a tensor are symmetric relative to one coordinate system, then this symmetry is preserved in all coordinate systems.*

DEFINITION. *A tensor is said to be skew symmetric if it changes sign every time two of its indices are interchanged.* Here it is assumed that we are dealing with a tensor all of whose indices are of the same nature, i.e., either all covariant or all contravariant.

The definition of a skew symmetric tensor implies that an even permutation of its indices leaves its components unchanged and an odd permutation multiplies them by -1.

The multilinear functions associated with skew symmetric tensors are themselves skew symmetric in the sense of the following definition:

DEFINITION. *A multilinear function* $l(\mathbf{x}, \mathbf{y}, \cdots)$ *of* p *vectors* $\mathbf{x}, \mathbf{y}, \cdots$ *in* $\mathbf{R}$ *is said to be skew symmetric if interchanging any pair of its vectors changes the sign of the function.*

For a multilinear function to be skew symmetric it is sufficient that the components of the associated tensor be skew symmetric relative to some coordinae system. This much is obvious from (9). On the other hand, skew symmetry of a multilinear function implies skew symmetry of the associated tensor (in any coordinate system). In other words, if the components of a tensor are skew symmetric in one coordinate system then they are skew symmetric in all coordinate systems, i.e., the tensor is skew symmetric.

We now count the number of independent components of a skew symmetric tensor. Thus let a_{ik} be a skew symmetric tensor of rank two. Then $a_{ik} = -a_{ki}$ so that the number of different components is $n(n-1)/2$. Similarly, the number of different components of a skew symmetric tensor a_{ijk} is $n(n-1)(n-2)/3!$ since components with repeated indices have the value zero and components which differ from one another only in the order of their indices can be expressed in terms of each other. More generally, the number of independent components of a skew symmetric tensor with k indices ($k \leq n$) is $\binom{n}{k}$. (There are no non zero skew symmetric tensors with more than n indices. This follows from the fact that a component with two or more repeated indices vanishes and $k > n$ implies that at least two of the indices of each component coincide.)

We consider in greater detail skew symmetric tensors with n indices. Since two sets of n different indices differ from one another in order alone, it follows that such a tensor has only one independent component. Consequently if $i_1, i_2, \cdots, i_n$ is any permutation of the integers $1, 2, \cdots, n$ and if we put $a_{12\cdots n} = a$, then

$$a_{i_1 i_2 \cdots i_n} = \pm a \tag{10}$$

depending on whether the permutation $i_1 i_2 \cdots i_n$ is even (+ sign) or odd($-$ sign).

EXERCISE. Show that as a result of a coordinate transformation the number $a_{12\cdots n} = a$ is multiplied by the determinant of the matrix associated with this coordinate transformation.

In view of formula (10) the multilinear function associated with a skew symmetric tensor with n indices has the form

$$l(\mathbf{x}, \mathbf{y}, \cdots, \mathbf{z}) = a_{i_1 i_2 \cdots i_n} \xi^{i_1} \eta^{i_2} \cdots \zeta^{i_n} = a \begin{vmatrix} \xi_1 & \xi_2 & \cdots & \xi_n \\ \eta_1 & \eta_2 & \cdots & \eta_n \\ \cdot & \cdot & \cdots & \cdot \\ \zeta_1 & \zeta_2 & \cdots & \zeta_n \end{vmatrix}.$$

This proves the fact that apart from a multiplicative constant the only skew symmetric multilinear function of n vectors in an n-dimensional vector space is the determinant of the coordinates of these vectors.

The operation of symmetrization. Given a tensor one can always construct another tensor symmetric with respect to a preassigned group of indices. This operation is called symmetrization and consists in the following.

Let the given tensor be $a_{i_1 i_2 \cdots i_n}$, say. To symmetrize it with respect to the first k indices, say, is to construct the tensor

$$a_{(i_1 i_2 \cdots i_k) i_{k+1} \cdots} = \frac{1}{k!} \sum a_{j_1 j_2 \cdots j_k i_{k+1} \cdots},$$

where the sum is taken over all permutations $j_1, j_2, \cdots, j_k$ of the indices $i_1, i_2, \cdots i_k$. For example

$$a_{(i_1 i_2)} = \tfrac{1}{2}(a_{i_1 i_2} + a_{i_2 i_1}).$$

The operation of alternation is analogous to the operation of symmetrization and permits us to construct from a given tensor another tensor skew symmetric with respect to a preassigned group of indices. The operation is defined by the equation

$$a_{[i_1 i_2 \cdots i_k]} = \frac{1}{k!} \sum \pm a_{j_1 j_2 \cdots j_k},$$

where the sum is taken over all permutations $j_1, j_2, \cdots, j_k$ of the indices $i_1, i_2, \cdots, i_k$ and the sign depends on the even or odd nature of the permutation involved. For instance

$$a_{[i_1 i_2]} = \tfrac{1}{2}(a_{i_1 i_2} - a_{i_2 i_1}).$$

The operation of alternation is indicated by the square bracket symbol [ ]. The brackets contains the indices involved in the operation of alternation.

Given k vectors $\xi^{i_1}, \eta^{i_2}, \cdots, \zeta^{i_k}$ we can construct their tensor product $a^{i_1 i_2 \cdots i_k} = \xi^{i_1} \eta^{i_2} \cdots \zeta^{i_k}$ and then alternate it to get $a^{[i_1 i_2 \cdots i_k]}$. It is easy to see that the components of this tensor are all kth order minors of the following matrix

$$\begin{bmatrix} \xi^1 & \xi^2 & \cdots & \xi^n \\ \eta^1 & \eta^2 & \cdots & \eta^n \\ \cdot & \cdot & \cdots & \cdot \\ \zeta^1 & \zeta^1 & \cdots & \xi^n \end{bmatrix}.$$

The tensor $a^{[i_i \cdots i_k]}$ does not change when we add to one of the vectors $\xi, \eta, \cdots$ any linear combination of the remaining vectors.

Consider a k-dimensional subspace of an n-dimensional space **R**. We wish to characterize this subspace by means of a system of numbers, i.e., we wish to coordinatize it.

A k-dimensional subspace is generated by k linearly independent vectors $\xi^{i_1}, \eta^{i_2}, \cdots, \zeta^{i_k}$. Different systems of k linearly independent vectors may generate the same subspace. However, it is easy to show (the proof is left to the reader) that if two such systems of vectors generate the same subspace, the tensors $a^{[i_1 i_2 \cdots i_k]}$ constructed from each of these systems differ by a non-zero multiplicative constant only.

Thus the skew symmetric tensor $a^{[i_1 i_2 \cdots i_k]}$ constructed on the generators $\xi^{i_1}, \eta^{i_2}, \cdots, \zeta^{i_k}$ of the subspace defines this subspace.

A CATALOG OF SELECTED

DOVER BOOKS

IN ALL FIELDS OF INTEREST

A CATALOG OF SELECTED DOVER BOOKS IN ALL FIELDS OF INTEREST

THE BOOK OF BEASTS: Being a Translation from a Latin Bestiary of the Twelfth Century, T. H. White. Wonderful catalog real and fanciful beasts: manticore, griffin, phoenix, amphivius, jaculus, many more. White's witty erudite commentary on scientific, historical aspects. Fascinating glimpse of medieval mind. Illustrated. 296pp. 5⅜ × 8¼. (Available in U.S. only) 24609-4 Pa. $5.95

FRANK LLOYD WRIGHT: ARCHITECTURE AND NATURE With 160 Illustrations, Donald Hoffmann. Profusely illustrated study of influence of nature—especially prairie—on Wright's designs for Fallingwater, Robie House, Guggenheim Museum, other masterpieces. 96pp. 9¼ × 10¾. 25098-9 Pa. $7.95

FRANK LLOYD WRIGHT'S FALLINGWATER, Donald Hoffmann. Wright's famous waterfall house: planning and construction of organic idea. History of site, owners, Wright's personal involvement. Photographs of various stages of building. Preface by Edgar Kaufmann, Jr. 100 illustrations. 112pp. 9¼ × 10.
23671-4 Pa. $7.95

YEARS WITH FRANK LLOYD WRIGHT: Apprentice to Genius, Edgar Tafel. Insightful memoir by a former apprentice presents a revealing portrait of Wright the man, the inspired teacher, the greatest American architect. 372 black-and-white illustrations. Preface. Index. vi + 228pp. 8¼ × 11. 24801-1 Pa. $9.95

THE STORY OF KING ARTHUR AND HIS KNIGHTS, Howard Pyle. Enchanting version of King Arthur fable has delighted generations with imaginative narratives of exciting adventures and unforgettable illustrations by the author. 41 illustrations. xviii + 313pp. 6⅛ × 9¼. 21445-1 Pa. $6.50

THE GODS OF THE EGYPTIANS, E. A. Wallis Budge. Thorough coverage of numerous gods of ancient Egypt by foremost Egyptologist. Information on evolution of cults, rites and gods; the cult of Osiris; the Book of the Dead and its rites; the sacred animals and birds; Heaven and Hell; and more. 956pp. 6⅛ × 9¼.
22055-9, 22056-7 Pa., Two-vol. set $20.00

A THEOLOGICO-POLITICAL TREATISE, Benedict Spinoza. Also contains unfinished *Political Treatise*. Great classic on religious liberty, theory of government on common consent. R. Elwes translation. Total of 421pp. 5⅜ × 8½.
20249-6 Pa. $6.95

INCIDENTS OF TRAVEL IN CENTRAL AMERICA, CHIAPAS, AND YUCATAN, John L. Stephens. Almost single-handed discovery of Maya culture; exploration of ruined cities, monuments, temples; customs of Indians. 115 drawings. 892pp. 5⅜ × 8½. 22404-X, 22405-8 Pa., Two-vol. set $15.90

LOS CAPRICHOS, Francisco Goya. 80 plates of wild, grotesque monsters and caricatures. Prado manuscript included. 183pp. 6⅝ × 9⅝. 22384-1 Pa. $4.95

AUTOBIOGRAPHY: The Story of My Experiments with Truth, Mohandas K. Gandhi. Not hagiography, but Gandhi in his own words. Boyhood, legal studies, purification, the growth of the Satyagraha (nonviolent protest) movement. Critical, inspiring work of the man who freed India. 480pp. 5⅜ × 8½. (Available in U.S. only)
24593-4 Pa. $6.95

ILLUSTRATED GUIDE TO SHAKER FURNITURE, Robert Meader. All furniture and appurtenances, with much on unknown local styles. 235 photos. 146pp. 9 × 12. 22819-3 Pa. $7.95

WHALE SHIPS AND WHALING: A Pictorial Survey, George Francis Dow. Over 200 vintage engravings, drawings, photographs of barks, brigs, cutters, other vessels. Also harpoons, lances, whaling guns, many other artifacts. Comprehensive text by foremost authority. 207 black-and-white illustrations. 288pp. 6 × 9. 24808-9 Pa. $8.95

THE BERTRAMS, Anthony Trollope. Powerful portrayal of blind self-will and thwarted ambition includes one of Trollope's most heartrending love stories. 497pp. 5⅜ × 8½. 25119-5 Pa. $8.95

ADVENTURES WITH A HAND LENS, Richard Headstrom. Clearly written guide to observing and studying flowers and grasses, fish scales, moth and insect wings, egg cases, buds, feathers, seeds, leaf scars, moss, molds, ferns, common crystals, etc.—all with an ordinary, inexpensive magnifying glass. 209 exact line drawings aid in your discoveries. 220pp. 5⅜ × 8½. 23330-8 Pa. $3.95

RODIN ON ART AND ARTISTS, Auguste Rodin. Great sculptor's candid, wide-ranging comments on meaning of art; great artists; relation of sculpture to poetry, painting, music; philosophy of life, more. 76 superb black-and-white illustrations of Rodin's sculpture, drawings and prints. 119pp. 8⅜ × 11¼. 24487-3 Pa. $6.95

FIFTY CLASSIC FRENCH FILMS, 1912–1982: A Pictorial Record, Anthony Slide. Memorable stills from Grand Illusion, Beauty and the Beast, Hiroshima, Mon Amour, many more. Credits, plot synopses, reviews, etc. 160pp. 8¼ × 11. 25256-6 Pa. $11.95

THE PRINCIPLES OF PSYCHOLOGY, William James. Famous long course complete, unabridged. Stream of thought, time perception, memory, experimental methods; great work decades ahead of its time. 94 figures. 1,391pp. 5⅜ × 8½. 20381-6, 20382-4 Pa., Two-vol. set $19.90

BODIES IN A BOOKSHOP, R. T. Campbell. Challenging mystery of blackmail and murder with ingenious plot and superbly drawn characters. In the best tradition of British suspense fiction. 192pp. 5⅜ × 8½. 24720-1 Pa. $3.95

CALLAS: PORTRAIT OF A PRIMA DONNA, George Jellinek. Renowned commentator on the musical scene chronicles incredible career and life of the most controversial, fascinating, influential operatic personality of our time. 64 black-and-white photographs. 416pp. 5⅜ × 8¼. 25047-4 Pa. $7.95

GEOMETRY, RELATIVITY AND THE FOURTH DIMENSION, Rudolph Rucker. Exposition of fourth dimension, concepts of relativity as Flatland characters continue adventures. Popular, easily followed yet accurate, profound. 141 illustrations. 133pp. 5⅜ × 8½. 23400-2 Pa. $3.95

HOUSEHOLD STORIES BY THE BROTHERS GRIMM, with pictures by Walter Crane. 53 classic stories—Rumpelstiltskin, Rapunzel, Hansel and Gretel, the Fisherman and his Wife, Snow White, Tom Thumb, Sleeping Beauty, Cinderella, and so much more—lavishly illustrated with original 19th century drawings. 114 illustrations. x + 269pp. 5⅜ × 8½. 21080-4 Pa. $4.50

SUNDIALS, Albert Waugh. Far and away the best, most thorough coverage of ideas, mathematics concerned, types, construction, adjusting anywhere. Over 100 illustrations. 230pp. 5⅜ × 8½. 22947-5 Pa. $4.50

PICTURE HISTORY OF THE NORMANDIE: With 190 Illustrations, Frank O. Braynard. Full story of legendary French ocean liner: Art Deco interiors, design innovations, furnishings, celebrities, maiden voyage, tragic fire, much more. Extensive text. 144pp. 8⅜ × 11¾. 25257-4 Pa. $9.95

THE FIRST AMERICAN COOKBOOK: A Facsimile of "American Cookery," 1796, Amelia Simmons. Facsimile of the first American-written cookbook published in the United States contains authentic recipes for colonial favorites—pumpkin pudding, winter squash pudding, spruce beer, Indian slapjacks, and more. Introductory Essay and Glossary of colonial cooking terms. 80pp. 5⅜ × 8½. 24710-4 Pa. $3.50

101 PUZZLES IN THOUGHT AND LOGIC, C. R. Wylie, Jr. Solve murders and robberies, find out which fishermen are liars, how a blind man could possibly identify a color—purely by your own reasoning! 107pp. 5⅜ × 8½. 20367-0 Pa. $2.50

THE BOOK OF WORLD-FAMOUS MUSIC—CLASSICAL, POPULAR AND FOLK, James J. Fuld. Revised and enlarged republication of landmark work in musico-bibliography. Full information about nearly 1,000 songs and compositions including first lines of music and lyrics. New supplement. Index. 800pp. 5⅜ × 8¼. 24857-7 Pa. $14.95

ANTHROPOLOGY AND MODERN LIFE, Franz Boas. Great anthropologist's classic treatise on race and culture. Introduction by Ruth Bunzel. Only inexpensive paperback edition. 255pp. 5⅜ × 8½. 25245-0 Pa. $5.95

THE TALE OF PETER RABBIT, Beatrix Potter. The inimitable Peter's terrifying adventure in Mr. McGregor's garden, with all 27 wonderful, full-color Potter illustrations. 55pp. 4¼ × 5½. (Available in U.S. only) 22827-4 Pa. $1.75

THREE PROPHETIC SCIENCE FICTION NOVELS, H. G. Wells. *When the Sleeper Wakes, A Story of the Days to Come* and *The Time Machine* (full version). 335pp. 5⅜ × 8½. (Available in U.S. only) 20605-X Pa. $5.95

APICIUS COOKERY AND DINING IN IMPERIAL ROME, edited and translated by Joseph Dommers Vehling. Oldest known cookbook in existence offers readers a clear picture of what foods Romans ate, how they prepared them, etc. 49 illustrations. 301pp. 6⅛ × 9¼. 23563-7 Pa. $6.50

SHAKESPEARE LEXICON AND QUOTATION DICTIONARY, Alexander Schmidt. Full definitions, locations, shades of meaning of every word in plays and poems. More than 50,000 exact quotations. 1,485pp. 6½ × 9¼. 22726-X, 22727-8 Pa., Two-vol. set $27.90

THE WORLD'S GREAT SPEECHES, edited by Lewis Copeland and Lawrence W. Lamm. Vast collection of 278 speeches from Greeks to 1970. Powerful and effective models; unique look at history. 842pp. 5⅜ × 8½. 20468-5 Pa. $11.95

PLANTS OF THE BIBLE, Harold N. Moldenke and Alma L. Moldenke. Standard reference to all 230 plants mentioned in Scriptures. Latin name, biblical reference, uses, modern identity, much more. Unsurpassed encyclopedic resource for scholars, botanists, nature lovers, students of Bible. Bibliography. Indexes. 123 black-and-white illustrations. 384pp. 6 × 9. 25069-5 Pa. $8.95

FAMOUS AMERICAN WOMEN: A Biographical Dictionary from Colonial Times to the Present, Robert McHenry, ed. From Pocahontas to Rosa Parks, 1,035 distinguished American women documented in separate biographical entries. Accurate, up-to-date data, numerous categories, spans 400 years. Indices. 493pp. 6½ × 9¼. 24523-3 Pa. $9.95

THE FABULOUS INTERIORS OF THE GREAT OCEAN LINERS IN HISTORIC PHOTOGRAPHS, William H. Miller, Jr. Some 200 superb photographs capture exquisite interiors of world's great "floating palaces"—1890's to 1980's: *Titanic, Ile de France, Queen Elizabeth, United States, Europa,* more. Approx. 200 black-and-white photographs. Captions. Text. Introduction. 160pp. 8⅞ × 11¾. 24756-2 Pa. $9.95

THE GREAT LUXURY LINERS, 1927–1954: A Photographic Record, William H. Miller, Jr. Nostalgic tribute to heyday of ocean liners. 186 photos of Ile de France, Normandie, Leviathan, Queen Elizabeth, United States, many others. Interior and exterior views. Introduction. Captions. 160pp. 9 × 12. 24056-8 Pa. $9.95

A NATURAL HISTORY OF THE DUCKS, John Charles Phillips. Great landmark of ornithology offers complete detailed coverage of nearly 200 species and subspecies of ducks: gadwall, sheldrake, merganser, pintail, many more. 74 full-color plates, 102 black-and-white. Bibliography. Total of 1,920pp. 8⅝ × 11¼. 25141-1, 25142-X Cloth. Two-vol. set $100.00

THE SEAWEED HANDBOOK: An Illustrated Guide to Seaweeds from North Carolina to Canada, Thomas F. Lee. Concise reference covers 78 species. Scientific and common names, habitat, distribution, more. Finding keys for easy identification. 224pp. 5⅜ × 8½. 25215-9 Pa. $5.95

THE TEN BOOKS OF ARCHITECTURE: The 1755 Leoni Edition, Leon Battista Alberti. Rare classic helped introduce the glories of ancient architecture to the Renaissance. 68 black-and-white plates. 336pp. 8⅜ × 11¼. 25239-6 Pa. $14.95

MISS MACKENZIE, Anthony Trollope. Minor masterpieces by Victorian master unmasks many truths about life in 19th-century England. First inexpensive edition in years. 392pp. 5⅜ × 8½. 25201-9 Pa. $7.95

THE RIME OF THE ANCIENT MARINER, Gustave Doré, Samuel Taylor Coleridge. Dramatic engravings considered by many to be his greatest work. The terrifying space of the open sea, the storms and whirlpools of an unknown ocean, the ice of Antarctica, more—all rendered in a powerful, chilling manner. Full text. 38 plates. 77pp. 9¼ × 12. 22305-1 Pa. $4.95

THE EXPEDITIONS OF ZEBULON MONTGOMERY PIKE, Zebulon Montgomery Pike. Fascinating first-hand accounts (1805–6) of exploration of Mississippi River, Indian wars, capture by Spanish dragoons, much more. 1,088pp. 5⅜ × 8½. 25254-X, 25255-8 Pa. Two-vol. set $23.90

DEGAS: An Intimate Portrait, Ambroise Vollard. Charming, anecdotal memoir by famous art dealer of one of the greatest 19th-century French painters. 14 black-and-white illustrations. Introduction by Harold L. Van Doren. 96pp. 5⅜ × 8½.
25131-4 Pa. $3.95

PERSONAL NARRATIVE OF A PILGRIMAGE TO ALMANDINAH AND MECCAH, Richard Burton. Great travel classic by remarkably colorful personality. Burton, disguised as a Moroccan, visited sacred shrines of Islam, narrowly escaping death. 47 illustrations. 959pp. 5⅜ × 8½. 21217-3, 21218-1 Pa., Two-vol. set $19.90

PHRASE AND WORD ORIGINS, A. H. Holt. Entertaining, reliable, modern study of more than 1,200 colorful words, phrases, origins and histories. Much unexpected information. 254pp. 5⅜ × 8½. 20758-7 Pa. $4.95

THE RED THUMB MARK, R. Austin Freeman. In this first Dr. Thorndyke case, the great scientific detective draws fascinating conclusions from the nature of a single fingerprint. Exciting story, authentic science. 320pp. 5⅜ × 8½. (Available in U.S. only) 25210-8 Pa. $5.95

AN EGYPTIAN HIEROGLYPHIC DICTIONARY, E. A. Wallis Budge. Monumental work containing about 25,000 words or terms that occur in texts ranging from 3000 B.C. to 600 A.D. Each entry consists of a transliteration of the word, the word in hieroglyphs, and the meaning in English. 1,314pp. 6⅞ × 10.
23615-3, 23616-1 Pa., Two-vol. set $27.90

THE COMPLEAT STRATEGYST: Being a Primer on the Theory of Games of Strategy, J. D. Williams. Highly entertaining classic describes, with many illustrated examples, how to select best strategies in conflict situations. Prefaces. Appendices. xvi + 268pp. 5⅜ × 8½. 25101-2 Pa. $5.95

THE ROAD TO OZ, L. Frank Baum. Dorothy meets the Shaggy Man, little Button-Bright and the Rainbow's beautiful daughter in this delightful trip to the magical Land of Oz. 272pp. 5⅜ × 8. 25208-6 Pa. $4.95

POINT AND LINE TO PLANE, Wassily Kandinsky. Seminal exposition of role of point, line, other elements in non-objective painting. Essential to understanding 20th-century art. 127 illustrations. 192pp. 6½ × 9¼. 23808-3 Pa. $4.50

LADY ANNA, Anthony Trollope. Moving chronicle of Countess Lovel's bitter struggle to win for herself and daughter Anna their rightful rank and fortune—perhaps at cost of sanity itself. 384pp. 5⅜ × 8½. 24669-8 Pa. $6.95

EGYPTIAN MAGIC, E. A. Wallis Budge. Sums up all that is known about magic in Ancient Egypt: the role of magic in controlling the gods, powerful amulets that warded off evil spirits, scarabs of immortality, use of wax images, formulas and spells, the secret name, much more. 253pp. 5⅜ × 8½. 22681-6 Pa. $4.00

THE DANCE OF SIVA, Ananda Coomaraswamy. Preeminent authority unfolds the vast metaphysic of India: the revelation of her art, conception of the universe, social organization, etc. 27 reproductions of art masterpieces. 192pp. 5⅜ × 8½.
24817-8 Pa. $5.95